SIMON FRASER UNIVERSITY
W.A.C. BENNETT LIBRARY

POLYMERIC FOAMS SERIES

POLYMERIC FOAMS

Science and Technology

POLYMERIC FOAMS

series editor Shau-Tarng Lee

INCLUDED TITLES

Polymeric Foams: Mechanisms and Materials
Edited by Shau-Tarng Lee and N. S. Ramesh

Thermoplastic Foam Processing: Principles and Development
Edited by Richard Gendron

Polymeric Foams: Science and Technology
Shau-Tarng Lee, Chul B. Park, and N.S. Ramesh

POLYMERIC FOAMS SERIES

POLYMERIC FOAMS

Science and Technology

Shau-Tarng Lee, Chul B. Park, and N.S. Ramesh

Taylor & Francis
Taylor & Francis Group
Boca Raton   London   New York

CRC is an imprint of the Taylor & Francis Group,
an informa business

Published in 2007 by
CRC Press
Taylor & Francis Group
6000 Broken Sound Parkway NW, Suite 300
Boca Raton, FL 33487-2742

International Standard Book Number-10: 0-8493-3075-0 (Hardcover)
International Standard Book Number-13: 978-0-8493-3075-9 (Hardcover)
Library of Congress Card Number 2006043863

Library of Congress Cataloging-in-Publication Data

Lee, S.-T. (Shau-Tarng), 1956-
  Polymeric foams : technology and science of polymeric foams / Shau-Tarng Lee, Chul B. Park, N.S. Ramesh.
        p. cm. -- (Polymeric foams)
  Includes bibliographical references and index.
  ISBN 0-8493-3075-0 (alk. paper)
  1. Plastic foams. I. Park, Chul B. II. Ramesh, N.S. (Natarajan S.) III. Title. IV. Polymeric foams series.

TP1183.F6L44 2006
668.4'93--dc22   2006043863

Taylor & Francis Group
is the Academic Division of Informa plc.

Visit the Taylor & Francis Web site at
http://www.taylorandfrancis.com

and the CRC Press Web site at
http://www.crcpress.com

Dedication

To the Lord, who gave us the spirit of power, of love, and of a sound mind.

Series Statement

The bubble is a wonderful creation: a perfect spherical shape, a beautiful arch in various degrees of curvature, and minimum surface area per given volume. Without the bubble, both art and science would definitely have a narrower scope. In fact, the bubble consists of a weak phase surrounded by and sustained in a strong phase. It is like a traditional Chinese virtue, Qian Xu (謙虛), or "empty inside" for receiving. Foam simply combines art, science, and philosophy, whereas, we admit, foaming could be one of the most mysterious phenomena in the universe. Fortunately, researchers and practitioners were able to turn that into advantageous technologies. Nowadays, foamed products are generally indispensable in our daily life.

It is known that foaming in the polymers involves delicate scientific mechanisms, subtle processing accuracies, unique morphology transformation and structure formation. It simply combines material principles, engineering designs, processing methodologies, and property characterization. Polymeric foams ride on the 20th century polymer industry high route to a fascinating success. Within a quarter of a century, the technology evolved from lab scale product to pilot line sample, then to commercial success. Today, it is viewed not only as a technique, but a well-established industry. Through challenges, such as ozone depletion, recycling and environmental regulation, in addition to upgrades, it became a strong industry.

Since polymeric foams encountered various upgrades—material/technology, emission/environment, property/application—it is crucial to maintain cohesiveness of polymeric foam by looking at it from various perspectives. This series is to cover material/mechanism, science/technologies, structure/property, application/post-usage, etc. The reader will gain an overall view as well as fascinating aspects of polymeric foam. We have to admit foaming is still mysterious in quite a few areas. It is my hope that a healthy and cohesive understanding can not only strengthen the structure of the existing polymeric foam industry, but generate further developments to reveal the basic truth. Let us not forget life and truth should go hand in hand.

S.T. Lee, Series editor
Sealed Air Corp., New Jersey

Preface

Polymeric foam exhibits enough extraordinary properties to differentiate it from the unique polymeric materials, which allows it to penetrate into almost all aspects of our daily life. Improvement of process technology, equipment and raw material availability made it possible to generate useful articles. Although the foam industry went through many challenges in the past decades it seems to be growing steadily due to better infrastructure, opportunities and global communication.

As we look back, it is amazing how polymeric foams have evolved from scientific concepts to lab research, then to pilot success, and finally to commercialization and a part of life in the last three quarters of a century. At present, foam researchers agree that foaming mechanisms remain somewhat mysterious, but the nucleation of spherical shape alone, a combination of science and art, is a fascinating truth. When a plethora of bubbles join together within a polymeric matrix to form a cellular structure, a collection of biblical principles, philosophy and sociology are nicely involved. It is important to understand that scientific research and application are two different areas, but they both are equally challenging and may share resources for breakthroughs.

Our attempt is very simple. Science and technology should be complementary to each other for a balanced and mutually beneficial growth. One without the other may cause an imbalanced outcome, and then self-destructive conflicts. When science and technology go hand in hand, they can withstand future challenges. From the 1930s to the 1950s, scientists laid the foundation of the foam industry, trailblazing foam technology in the 1960s through the 1980s. Since then, teamwork was acknowledged in dealing with ozone and application issues. Regulation and performance continue to be the main driving force for the global foam industry. From the Montreal protocol to the Kyoto protocol, the challenge of the global foam industry becomes even greater. A healthy future must be based on a solid science and technology foundation.

This book offers a clear guideline to link the basic science and foaming technologies. The first three chapters of this book cover the scientific principles and fundamental foaming mechanisms. The next three chapters are dedicated to general foaming technologies and product applications. Chapters 7 and 8 cover recent developments in the composite and degradable fields, which not only serve as current examples of using mechanisms and technologies to meet application needs, but show the role that polymeric foam is playing in the future developments of the global material realm. This book could be used as a supplementary book for seniors in chemical engineering, mechanical engineering, and material science. For polymer science and engineering, this book can be considered as a co-text book for graduate school.

S.T. Lee

C.B. Park

N.S. Ramesh

Acknowledgments

This book is a wonderful example of teamwork and learning experience. Communication, collaboration, and commitment were progressively improved in the last 4 years and are still improving. Although, frustrations, setbacks, and dead ends were inevitable; however, good eventually overcame evil; joy replaced suppression, and team effort overcame individual work. We thank God for the talents, support, cooperation, friendship, patience, perseverance, the ability to deal with uncertainties, and much more.

We acknowledge permission and support from Sealed Air Corporation and University of Toronto to allow the work to go to the public. The principal author is very thankful for the quality review comments from Dr. Masayuki Yamaguchi of Japan Advanced Institute of Sciences and Technology, Professor Masahiro Ohshima of Kyoto University, and Dr. Michel Huneault of Canadian Research Council, and the valuable marketing information from David Rives of Sealed Air Corporation. The extra hours from Hyun-Jung (Jenny) Jun, Kevin L. Lee, Slavec Kubicz, Arlene LaDuca, Nancy Demains at Sealed Air Corporation and from Guanming Li, Remon Pop-Iliev, and Gangjian Guo at University of Toronto are deeply appreciated. Special appreciation goes to Leonard Kareko, whose diligence and dedication in the final stages of this book greatly helped the delivery of the undertaking. Without their faithful dedication in making drawings, scanning charts, checking references, designing the cover, formatting manuscripts and making revisions, we most certainly would have compromised the quality of the book and delayed printing.

Our gratitude goes to our wives: Mrs. Mindy Park, Mrs. Malathi Ramesh, and Mrs. Mjau-Lin Lee. Their unconditional support and willingness to endure the somewhat rough preparation process are especially acknowledged. We also acknowledge our gratitude to all our family members for their love and for providing us with the best education since our childhood days. May God use this book to enlighten the readers, and may their growth become our reward and a benefit to the next generation in which the truth will continue to expand and to shine.

Authors

Shau-Tarng Lee, who was born and raised in Taipei, Taiwan, ROC, received his Bachelor of Engineering degree from the National Tsing-Hua University. He joined graduate school in the Chemical Engineering department at Stevens Institute of Technology in 1980 under Professor Joseph Biesenberger's guidance in foam enhanced devolatilization, and, in 1981 and 1982, spent summer internships with Farrel Company to investigate bubble phenomena in devolatilization. He also received a Stanley fellowship and grant from the National Science Foundation (NSF) to support his research works at Stevens. After earning his master's degree in Engineering and his Ph.D., he joined Sealed Air Corporation in 1986. Since then, he has specialized in foam extrusion research, development, and production support as Development Engineer, Assistant Research Director, and presently Research Director.

Dr. Lee has accomplished more than 100 publications, including 26 U.S. patents, and was elected to the Fellow of Society of Plastics Engineers in 2001. He was inducted to Sealed Air's Inventor Hall of Fame in 2003. He is the editor for *Foam Extrusion; Principles and Practice*, published by Technomic Publishing Co. (now Taylor & Francis) in July, 2000, and is also Polymeric Foam Series editor for CRC Press (now Taylor & Francis), with two volumes published in 2004: *Mechanisms and Materials* (edited by S.T. Lee and N.S. Ramesh), and *Thermoplastic Foam Processing* (edited by R. Gendron). Dr. Lee and his wife, Mjau-Lin Tsai, have three children, Joseph, Matthew, and Thomas. Currently, they reside in Oakland, New Jersey. He is a local church elder, and is actively involved in mission works in Asia.

Chul B. Park received his Ph.D. from MIT in 1993. He is a professor and holder of the Canada Research Chair Tier I in Advanced Polymer Processing Technologies at the University of Toronto. He is also the founder and director of the Microcellular Plastics Manufacturing Laboratory, which enjoys the reputation of being one of the pioneering research institutions in refining various foaming technologies. As a Fellow of SPE, ASME, and CSME, Dr. Park is an accomplished scientist with international recognition in the field of polymer foam processing. He is the author or co-author of more than 300 publications including 15 patents and 110 journal papers. He is also active in professional activities. He is the editor-in-chief of the *Journal of Cellular Plastics* and serves as an advisory editorial board member of *Cellular Polymers and Advances in Polymer Technology*. As the Technical Program Chair, Dr. Park has been organizing the Foam Symposiums at PPS and the program of Foams TopCon 2006. He also serves as an active board member of the Thermoplastics and Foams Division of SPE. Currently, he resides with his wife Mindy, his son Joshua and his daughter Esther in Toronto, Ontario.

N.S. Ramesh is the Director of R&D of Specialty Materials Business at Sealed Air Corporation. He received his bachelor's degree in Chemical Engineering from the University of Madras, India and has master's and Ph.D. in Chemical Engineering from Clarkson University in Potsdam, New York. He attended the Summer Rheology Program at MIT and the executive management programs at University of Michigan (Ann Arbor) and SMU Cox (Dallas) business schools. Dr. Ramesh has worked in the area of polymeric foams and materials for 17 years and received Best Paper awards twice for his pioneering work on foams from the Society of Plastics Engineers (SPE). He has more than 50 publications, including 22 U.S. patents, two books, and two book chapters. The SPE elected him a Fellow of the Society in 2002 and he served as Technical Chair for the SPE Foams Conference from 1998–2004. He was inducted into Sealed Air's Hall of Fame in March 2006. His wife, Malathi Ramesh, teaches elementary school children; his oldest son, Deepak Ramesh, is a student at Rice University, and his second son, Vijay Ramesh, is a fourth grader. All are involved with local church and voluntary outreach activities. Without the support of his immediate family and parents, this book would not have been possible.

Table of Contents

1

Introduction to Polymeric Foams

1.1 Basic Considerations on Foams, Foaming, and Foamed Polymers

Matter generally assumes one of three forms (or phases): gas, liquid, or solid. Gases are essentially shapeless and formless, and naturally or artificially exist or co-exist with the other two phases, such as in sponge, cork, aerogel, cake, for example. In fact, gas molecules are capable of penetrating into a liquid or solid phase to create mixtures. Table 1.1 presents a summary of typical gas/liquid and gas/solid mixtures.

The word "foam" derives from the medieval German word *veim* for "froth" [1]. "Foam" refers to spherical gaseous voids dispersed in a dense continuum. There are a number of common natural and artificial foamed products, ranging from foamed pumice to seat cushions [2,3] (see Table 1.2).

TABLE 1.1

Typical Two-Phase Systems

Phenomena	Terminology
Gas bubbles on top of liquid	Froth
Gas bubbles dispersed in liquid	Emulsion Bubble
Liquid bubbles in liquid	Emulsion Liquid
Liquid bubbles in solid	Jelation
Gas Bubbles in solid	Foam

TABLE 1.2

Common Foamed Products

Natural:	Pumice, Tree Trunk, Wood, Cork, Marine Organisms
Synthetic:	
Food:	Steamed Rice, Flour Dough, Popped Cereal…
Plastic:	Seat Cushion, Life Jacket, Insulation Board…
Automotive:	Arm-rest, Liner, Bumper…
Sports:	Helmet Pad, Knee Protection, Surfing Board
Medical:	Tape, Gasket Seal…

Foaming occurs when free gas molecules are converted into spherical bubbles, and typically takes place when the surrounding conditions change too abruptly to allow a smooth response from the system. Although, at modest changes, molecular diffusion may be adequate to restore the state of equilibrium, drastic changes usually preclude reaching the equilibrium through conventional transport mechanisms, such as diffusion and vaporization. Hence, foaming can be understood as a way of dissipating the disturbances in a given environment. However, foaming can also be regarded as a transition from a stable (homogeneous) state to a meta-stable or unstable (heterogeneous) state [4]. In the case of boiling, for example, under heating the stable liquid becomes the unstable bubbling which will stop when the liquid is exhausted to form a homogenized vapor state.

The dense phase surrounding the gas bubbles can be further strengthened (e.g., by cooling) to retain the useful effects of the dynamically intensive foaming process and thereby deliver stable foamed products. The hardening process must proceed faster than the condensation of the gas phase otherwise bubbles may shrink back to liquid state. In certain cases, the surrounding may be inadequate to hold the shrinkage-induced vacuum. It tends to collapse back to the unfoamed state. In other words, timing is very important. Foaming is thus a very unique technology, turning unstable foaming into a stable and useful product.

The foaming methodology usually consists of introducing a gaseous phase into a melt, then foaming the gas, and subsequently solidifying the melt before gaseous bubbles condense or collapse back to a liquid state. Gas bubbles are generated in a spherical shape by virtue of either entrainment or nucleation. Since the spherical form has the lowest surface energy for a given volume, it is the ideal shape for the weak (gaseous) phase to sustain within a dense (liquid or solid) phase. As illustrated in Figure 1.1, the gas is considered to be the weak phase, and therefore it has to counterbalance the summation of both the surrounding pressure and structural force, F, in order to survive in the form of a bubble. Thus:

$$P_B = P + F/A \qquad (1.1)$$

where, P_B, P, and A, denote the bubble pressure, surrounding pressure, and bubble surface area, respectively. When the surrounding phase is in a liquid state, Equation (1.1) becomes:

$$P_B = P + 2\sigma/R_B \qquad (1.2)$$

where, σ and R_B represent the surface tension and bubble radius, respectively.

As Table 1.3 makes evident, there are various possible ways of introducing external gas molecules into a given volume occupied by liquids or solids. When gas exists in a dense matrix, the gas molecule is small enough to establish itself in structural holes of the surrounding liquid or solid.

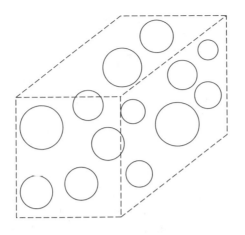

FIGURE 1.1
Gaseous bubbles in a dense matrix.

TABLE 1.3

Methods of Implementing Gas into Liquid and Solid

Dissolve gas by pressure and/or mixing into solid and liquid
Blow gas into melt, such as: gas-assisted injection molding
Chemical reaction to evolve gas in liquid or melt
Partial decomposition of molten polymer
Blend encapsulated gas phase in polymer into solid or liquid for further expansion
  in processing
Reaction with liquid to form bubble, such as: sodium bicarbonate in water
Entrain air into liquid for high-temperature or low-pressure (vacuum) processing

However, for a bubble to be produced, a sufficient number of gas molecules must gather in order to overcome the resistance of the surrounding matrix. In this context, an important parameter known as the critical bubble radius, can be determined according to the following equation:

$$R_{cr} = 2\sigma/(P_B - P) \tag{1.3}$$

Hence, if the bubble radius is smaller than the critical radius, we can infer that the surface tension force is high enough to cause the gas clusters to collapse. In a state of ultimate equilibrium, P_B should be the same as P, and therefore, R_{cr} becomes an infinity. This indicates that a spherical shape would not be achieved because any bubble size is smaller than the infinite critical radius. But when the equilibrium is destroyed, the system will go through a series of non-equilibrium states to reach another equilibrium, and in this process, bubbling may occur. For instance, when a lower pressure is suddenly applied (i.e., $P < P_B$) to an equilibrium condition, bubbling becomes a way to re-establish another equilibrium. Bubbling is an aggressive way of generating a significant amount of interfacial area for diffusion, which in turn dissipates any energetic inequalities. This dynamic phenomenon in an unstable state is very natural and continual bubbling (e.g., boiling), and bubble growth (e.g., polymeric foaming) will occur until the system becomes stabilized. At any event, this dynamic state can be timely harnessed for useful applications.

It should be noted here that when the bubble is very tiny in size, the bubble pressure is relatively high in order to sustain the surface tension force. Although the buoyancy force is low because of the small bubble size, the high internal bubble pressure renders the foaming process highly unstable. These unstable bubbles expand in size, which in turn dissipates the pressure gradient across the interface. If the surrounding resistance is low, buoyancy becomes manifest, like when the bubbles in soda or in soapy water are collected on top of the liquid. The rupture of the bubble walls would be inevitable as the bubbles meet and the thinning of the bubble walls progressively occurs because of the material drainage in the cell walls. In a dense syrup, bubble movement is stifled by the surrounding viscous phase. As a consequence, bubbles appear to have a much longer lifespan, creating a greater opportunity for the viscous phase to solidify into a "permanent" gas/solid structure.

Polymers represent a very special group of materials which are quite different from, for example, closely packed materials, such as glass, metal, rock, or wood. "Poly" in Greek means "multitude," while the suffix "mer" signifies unit (i.e., monomer). In fact, a polymer consists of many long-chain polymers entangled and/or bonded by functional groups and van der Waals forces. Conversely, polymers can be made by prompting reactions between functional groups or by free particle addition to form a long-chain polymer. The former type of polymer normally possesses a network structure, which foams after a heat-induced reaction is complete. As a result, its fluidity would no longer be thermally responsive, especially once a 3-D structure is formed. By contrast, the latter kind of polymer changes from a solid to a molten state, and vice versa, in response to temperature changes. Thus, polymers (or plastics) are categorized as thermoplastics or thermosets according to the way they respond to thermal variations.

If thermosets are somehow exposed to elevated temperatures, their polymer chains become susceptible to forming a network structure; they are then no longer thermo-reversible. However, when polymers possess various chain lengths held together by van der Vaal forces and chain entanglements, they demonstrate a unique visco-elastic behavior, as illustrated in Figure 1.2.

Unlike crystals, which are characterized by a sharp change of state occurring at a given melting point, thermoplastics convert from a solid to a liquid state within a specific temperature window. Below the glass-transition temperature, T_g, polymer chains are basically frozen. As the temperature exceeds T_g, chain movement begins to increase. If a polymer possesses crystallinity, it tends to melt at the melting point to form a polymeric melt. Because of their thermo-reversible nature, thermoplastics are the preferred class of materials for processing purposes. In the molten state, they can blend and

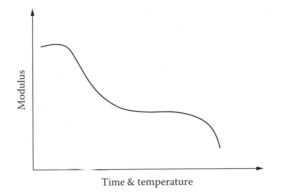

FIGURE 1.2
Viscoelastic nature of Polymer, variation of elastic in time and temperature domain.

co-exist with other polymers or gases. Moreover, the visco-elastic nature of thermoplastics endows them with the appropriate strength (or resistance) to control the unstable foaming phenomena and thereby facilitate a plausible path to reach a stable foamed plastic product [5].

It is important to note that thermoplastic and thermoset foams are manufactured according to very different processes. The successful development of thermoplastic foams is strongly dependent on the kind of equipment that determines the capacity to process the thermoplastic having unique thermally reversible characteristics with a gas that is introduced. Essentially, foaming is a pressure- and temperature-controlled phenomenon. On the other hand, the production of thermoset foams seems to hinge more on the materials "formula" and their reaction kinetics. Table 1.4 compares thermosets and thermoplastics, and their respective foaming mechanisms [6].

Foaming is a unique phenomenon that can be effectively used in a number of applications. For example, the tremendously increased surface area in foams makes them suitable for effective mass transfer in separation operations such as foam-enhanced devolatilization [7]. In another example, a separation tower can literally be reduced to a unit operation in a plasticator where a foamed structure can be induced and destroyed to achieve the removal of the undesired volatile contents from the polymeric melt. Another example of a natural foam at work is the human lung, in which cellular tissues exist to enhance gas exchange. The cell wall strength substantially

TABLE 1.4

Foaming Mechanism Comparison between Thermoplastic and Thermoset Foam

	Gas within System	Foaming	Stabilization
Thermoplastics	Dissolution or Decomposition	Supersaturation	Cooling
Thermoset	Chemical Reaction	Gas Evolution	Polymerization, X-linking and Cooling

affects the efficiency of gas exchange in the pulmonary tissue. By extension, exercising our muscles can improve tissue strength and thus enhance the healthy transfer of gas [8].

However, the presence of gaseous voids in the product will reduce the heat transfer rate due to the high insulation properties of the foamed structure. This can cause serious concerns about the safety on account of reaction heat accumulation [9]. The presence of voids can also be a disadvantage from an aesthetics and/or mechanical strength point of view.

1.2   History of Polymeric Foams

Over the years, the foam processing of plastics has become the subject of much discussion, attracting extensive and intensive research and development efforts within the disciplines of chemistry, material and engineering [10–13]. A plastic itself possesses a unique property/weight ratio, and its thermal dependency makes it quite different from other materials. It is no wonder that plastics constitute their own distinct material industry. Foamed plastics indeed represent an important extension of the polymer property spectrum and have an enhanced property/weight ratio, therefore offering some unique advantages [14].

The production of both thermoplastic and thermosetting materials has contributed appreciably to the development of the polymeric foam industry, which itself dates back to the first half of the 20th century. It used to be the case that some polymers would be virtually tested with a gas in laboratory environments, and then implemented in pilot- and commercial-scale foaming processes. However, the unique properties of foams have opened up the plastics industry to a wide variety of possible applications; this has been the driving force behind the accelerated development of foaming technologies during the second half of the 20th century.

The development of various technologies for polymer synthesis and more recently, of newly designed polymer processing equipment, was the key factor that propelled the development of polymeric foams between the 1950s and the 1970s. Based on this infrastructure, dedicated efforts from scientists and engineers around the world resulted in an increased understanding of foaming mechanisms and in enhanced techniques for efficient foam production. After the 1980s, increasing insight in environmental issues of both polymeric materials and blowing agent contributed further to the reinforcement of the foam industry. Although it still appears diversified, significant efforts have been made in the industry towards bridging the gap between a purely scientific and a purely practical approach [14,15]. Table 1.5 presents a number of common methods for making polymeric foams [15–17], whereas Table 1.6 highlights the crucial milestones in the development of polymeric foams [18–31].

TABLE 1.5

Polymeric Foam Methodology

Thermoplastic:	Extrusion Foaming, Injection Molding Foaming, Bead Foaming, Rotational Molding Foaming, Compression Molding Foaming, Oven Heat Foaming, Coex Foaming
Thermoset:	Reactive Foaming, Reactive Mold Foaming, Reactive (or Reaction) Injection Molding (RIM) Foaming

TABLE 1.6

Highlights of Polymeric Foam Developments

Time	Contents	Authors or Companies	References
1931	Foamed Polystyrene	Munters and Tandberg	U.S. Pat. 2,023,204 [18]
1937	Foamed Polyurethane (PU)	Dr. Otto Bayer	K.C. Frisch [19]
1941	Foamed Polyethylene	Johnson, F. L.	U.S. Pat. 2,256,483 [20]
1944	Extruded Polystyrene Foam	Dow Chemical	[21]
1945	Rigid PU Foam	Germany	PU at Farben, Report 1122 [22]
1952	Flexible PU Foam	Germany	K. C. Frisch [19]
1954	Expandable Bead	Stastney and Goeth	U.S. Pat. 2,681,321 [23]
1959	Rigid PU Foam Produce	ICI	G. Woods [16]
1962	PS Foam Injection Molding	Beyer et al.	U.S. Pat. 3,058,161 [24]
1962	Extruded Ethylene Foam	Rubens et al.	U.S. pat. 3,067,147 [25]
1967	Twin Screw for Foam Brt. Pat. 1,152,306	Spa, L. M. P.	It. Pat. 795,393 [26]
1967	ABS foam; Injection Mold	Woollard, D.	SPI 12th Ann. Conf. [27]
1968	Rigid Isocyanurate Foam	ICI	G. Woods [16]
1972	Extruded Propylene Foam	Parrish, R. G. (DuPont)	U.S. Pat. 3,637,458 [28]
1982	Accumulator Extrusion	Collins, F. (Valcour)	U.S. Pat. 4,323,529 [29]
1984	PP Molded Foam Article	Japan Styrene Paper	Jap. Pat. 59-23731 [30]
1990	PET Foam Extrusion	Shell/Petlite®	Xanthos, D. 2000 [31]

The steady growth of polymeric foam consumption in the last several decades is solid evidence of the importance of foam to our society. Foam consumption surpassed metal usage in the mid 1980s, and similarly, plastics have replaced numerous traditionally wooden products. Figure 1.3 shows the property ranges for different materials used in the products [32,33].

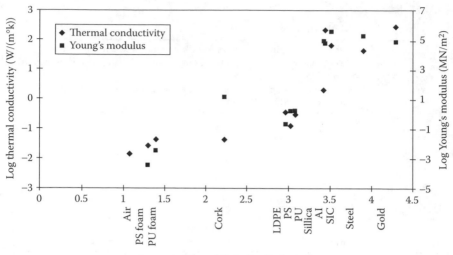

FIGURE 1.3
Variation of thermal conductivity and Young's modulus for different materials in log-log scale with density as the independent ordinate (data collected from [32] and [33]).

1.3   Foam Structure and Properties

A polymeric foam possesses unique physical, mechanical, and thermal properties, which are governed by the polymer matrix, the cellular structure and the gas composition. When a gaseous phase is being dispersed in a spherical form within a polymeric matrix, a composite structure is naturally formed, and the properties of this composite are determined by its constituents and their distributions. Since the weight of gas is negligible, the properties of gas/polymer composite often volumetrically depend on the participating components. The density is a typical example of this, especially in cases where the bubble phase dominates. However, the thermodynamic properties such as the specific heat, the equilibrium constant, and the heat conductivity, would still remain gravimetrically dependent on the individual elements, i.e., by weight of each element.

The characteristics of polymeric foams are determined by the following structural parameters: cell density, expansion ratio, cell size distribution, open-cell content, and cell integrity. These cellular structural parameters are governed by the foaming technology used in processing, and the foaming technology often heavily depends on the type of polymer to be foamed. In other words, different polymers display different properties, and thus distinct processing systems are required to accommodate these discrepancies [34,35]. This is why various foaming technologies, such as batch

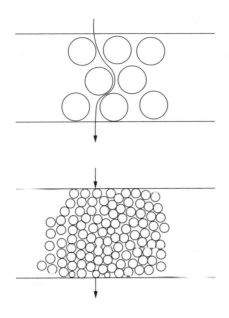

FIGURE 1.4
Heat and mass transfer across the polymeric foam at different cell size distributions.

(single-stage, multi-stage), semi-continuous, and continuous processes, have been gradually developed over the years for each specific polymer foam. Different polymer foams exhibit different properties depending on the characteristics of foams which can range from soft to stiff modulus, resilient to tough behavior, low to high hysterisis, and mono to multi-model cell distribution. In this context, it would be desirable to discuss the fundamental relationships between the structure and the properties before discussing the foaming methodology and application diversification.

When the bubble phase is limited, the polymer mainly dictates the properties of the foam. As illustrated in Figure 1.4, the bubble's existence detours the direction of the heat transfer because the convection heat transfer in the gas phase of the bubble is negligible. However, when the bubble phase dominates (as in the case of a highly expanded foam), the polymer's ability to hold the gas within the integral cell is higher, and yet, the polymer contributes less to the foam's final properties. Therefore, in many instances, the contributions of the gas phase could be dominant in determining the properties of the foams. For example, the energy absorption ability of low-density polyethylene foams is governed by the gaseous phase. Namely, the dispersion of many grown cells results in a superior absorption capability when compared with the same amount of gas encapsulated in a polymeric film, such as in a sealed bag or pillow.

Not only are the amount and distribution of cells critical parameters in establishing the final property configuration of a given polymeric foam, but the nature of the cells (open cell versus closed cell) similarly plays an important role in determining the properties, which, in turn, dictate the foam's possible applications. When the integrity of the cell wall is poor and does not hold the gaseous phase together, it instead prompts interconnections with the neighboring cell(s). Therefore, under deformation, the gas compression, which is a key component dictating the mechanical response of the cell, may no longer be effective. Despite an eventual quick recovery, after the deformation has been released, both the compressive strength and the energy absorption ability would be weakened. Other mechanical properties would also be adversely affected. However, open cells have capillaries that allow for unique fluid absorption features; this is a desirable feature, for instance, in the wet meat industry. Open cell foams also perform well in sound-deadening applications, where sound waves attenuate after numerous bounces.

On the other hand, the cell size exerts a great impact in disturbance distribution. According to the studies of the microcellular foams with a cell size in the order of 10 microns, small cells provide an improved energy absorption capability, which will be further described in a later section. It is well known that the insulation ability of polyurethane and polystyrene foams strongly depends on the cell size. The smaller the size, the higher the insulation, partly because of the reduced radiation effects in the cell during heat transfer [36]. The other vital consideration is the open-cell content: it seems to decrease as the cell size decreases. Nonetheless, when large foam expansion occurs (e.g., over 10 times), the cell morphology greatly contributes to the overall thermal and mechanical performance of the foams.

The amount of residual blowing agent in the cells affects the insulating property of the polyurethane foam and the extruded or bead polystyrene insulation foams. Because of the low diffusivity of HCFCs and HFCs in the polymeric matrix, a blowing agent with a high insulating property does not diffuse out quickly and any remaining blowing agent in the cells and cell structures will determine the overall insulating property of the foam. Since the blowing agent eventually is replaced by air (as described in the next section), the insulating property of the polymeric foam decreases with time.

1.4   Blowing Agents

Any gas can serve as a blowing agent, but not every gas can be easily implemented in the foaming process. In fact, some gases are friendlier than others, in terms of solubility, volatility, and diffusivity. It appears that the blowing agent's quality, quantity, and nature are key factors in the production of a foamed structure with certain desired properties. The blowing agent governs the selection of the foaming methodology, which more often than not becomes a limiting factor in industrial practice. It should be noted here that most blowing agents are easier to introduce into a polymer in any foaming equipment than air although the foam will consist of polymer and

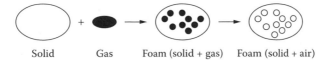

Solid   Gas   Foam (solid + gas)   Foam (solid + air)

FIGURE 1.5
General thermoplastic foaming path; gas from without to within the polymer, eventually re-placed by air.

air ultimately. Once the foam is formed and exposed to air, most blowing agents are replaced by air as time passes. In most polymeric foams, the role of the blowing agent is like that of a "catalyst"—it participates in the process (see Figure 1.5). From the perspective of the blowing agent, the foaming process can be seen as a succession of three steps: implementation, liberation, and evacuation. In simple terms, (i) implementation refers to the process used for introducing predetermined quantities of an external gas into the polymer matrix to form a polymer/gas solution; (ii) liberation suggests the conversion of the polymer/gas solution, which is characterized by non-differentiated (invisible) structural elements, into a fully-differentiated (visible) cellular structure; and (iii) evacuation signals the transformation of the polymer foam from a blowing-agent-filled to an air-filled state. The respective key mechanisms are illustrated in Figure 1.6.

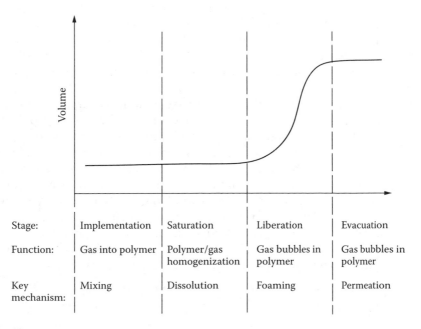

Stage:	Implementation	Saturation	Liberation	Evacuation
Function:	Gas into polymer	Polymer/gas homogenization	Gas bubbles in polymer	Gas bubbles in polymer
Key mechanism:	Mixing	Dissolution	Foaming	Permeation

FIGURE 1.6
Polymer volume variation as foaming progresses; function and key mechanism attached.

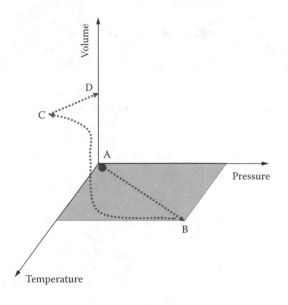

FIGURE 1.7
P-T-V change of foam extrusion from A to D; A: low V, low P, low T., B: low V, high P, high T, C: high V, low P, high T, D: high V, low P low T.

The process of thermoplastic foaming can be characterized by state change (see Figure 1.7). Once the raw plastic material (State A) is heated and pressurized, a blowing agent is added (State B). The foam structure is developed by lowering the pressure (State C), and finally, a foam product is yielded by cooling the polymer matrix. It is the blowing agent that plays a major role in changing these states.

The fundamental features of a blowing agent are its capability to migrate throughout the polymer, out of the polymer, and away from the polymer. In other words, the thermodynamic, kinetic, and transport properties of the gas are of vital importance at different stages of the process. Therefore, a balanced view should be used in analyzing and selecting a proper blowing agent in order to produce a desired foam structure out of a polymer. In general, chemical reactions and physical mixing are the main methods of introducing a blowing agent into the polymeric melt. The former implies creating a reaction-induced gas in the solution, whereas the latter assumes gas dissolution into the polymeric melt. Table 1.7 presents the key properties that should be considered when selecting a blowing agent.

TABLE 1.7

Blowing Agent Property Considerations

Aspect	Properties
Thermodynamic properties	Volatility, Solubility, Molecular Weight, Equilibrium Constant, Surface Tension, Boiling Point, Specific Volume at STP
Transport Properties	Diffusivity, Permeability
Processing and Environment	Plasticization, Flammability, Stability, Toxicity, Ozone, Warming
Application:	Odor, FDA, Conductivity

1.5   Environmental Issues and Technical Challenges

In 1927, Midgley succeeded in synthesizing halogenated carbon compounds, or what we now commonly refer to as chlorofluorocarbons (CFCs), to replace action/ammonia as refrigerants in Europe. Since then the non-toxic, non-flammable, and stable natures of CFCs have attracted wide and deep interest from various industries [37–39]. In 1988, the global annual production of CFCs exceeded ten billion pounds; a very substantial portion of this amount was used for foam production. However, the stable CFCs tend to release chlorine under ultraviolet (UV) exposure in the high atmosphere, which in turn retards the ozone formation reaction. As a result of the environmental threat posed by CFCs, the Montreal protocol was established in 1987 in an effort to officially enforce the emission control and usage of CFCs. In addition, heavy taxes were proposed and put into effect to be associated with CFC purchase to reduce consumption. In the early 1990s, Hydrochlorofluorocarbons (HCFCs) were first introduced as a replacement for CFCs. However, according to scientific findings, HCFCs also released chlorine, which continued to contribute to ozone deterioration. At the same time, the global warming phenomenon was receiving increased attention. Also known as "the green-house effect," global warming is the consequence of slowing down the dissipation of heat from the earth's surface. Earth scientists and meteorologists predict that the earth's surface temperature will continue to increase over the next centuries. In consequence, new or existing blowing agents must conform to environmentally sound standards, and it was finally decided to include HCFCs on the phase-out list.

The inclusion of a large molecular weight physical blowing agent with a low diffusivity is necessary in the production of low density (high expansion) polymeric foams. This is especially true in the case of thermoplastic foams. Since the previously used CFC-based and HCFC-based blowing agents are environmentally hazardous, new blowing agents have been searched. However, the development of a reliable and safe alternative blowing agent should

satisfy both environmental concerns as well as issues of product performance. Although the polyurethane foaming process creates a sufficient
amount of blowing gases, most of these gases possess a higher thermal
conductivity than CFCs, and provide either a low-energy efficiency or a
bulkier unit. Therefore, the present search for new physical blowing agents
with certain desired properties (see the list of properties in Table 1.7) continues. Currently, it appears that hyrdrofluorocarbon (HFC), hydrocarbon
(HC), and their blends are the main candidates for the new blowing agent.
HCs are readily available from petroleum cracking and are very compatible
with most polymers, however they are flammable and create smog (smoke
+ fog) under the influence of UV. On the other hand, the future of HFCs is
dependent upon their unknown effects on the environment and their cost
required to produce the foams having the desired structures and properties.
Any new scientific findings about the potential effects of the blowing agents
may be critical to new regulations. But despite the great challenge, it is
encouraging to note that the industrial companies in the global foam industry have agreed to change the blowing agents to protect the ozone layer.
Also, it is worth noting that the foam industry has grown stronger technically
in the last decade, as a result of the technical challenge of replacing the
blowing agent. Especially, some major suppliers of blowing agents have
committed to the manufacture of HFCs in the early 2000s [40,41]. In addition
to the environmental issues of the blowing agent, a better understanding of
polymeric foam is necessary in order to respond to new possible regulatory
challenges.

Thermoplastic foams can also be looked at from the perspective of the
state and composition of gas and polymer. Their processing path is illustrated in Figure 1.8, where one can observe how the blowing agent ultimately
behaves as a catalyst as discussed earlier. The thermoplastic foam undergoes
an aging process that will last depending on the dimension of the foamed
plastics; it may take anywhere from a few seconds for a millimeter thick
sheet to a few months for a centimeter thick sheet. The final polymeric foam
product is comprised of polymer and air. The question as to why compressed
air or nitrogen is not used as potential blowing agents naturally arises. The
reasons for this lie in the fact that one of the major components of air, O_2,
causes oxidation of the polymer. Furthermore, the low solubility of nitrogen
presents a real challenge for foam processors, although nitrogen has been
used in a small section of foam production for low expansion foams.

After numerous years of development, the manufacture of polymeric
foams appears to be moving in two independent directions according to two
different considerations: material saving and performance. Both factors are
governed by the performance/weight (or performance/cost) ratio. In the
case of microcellular materials, material savings would evidently serve two
purposes: mass conservation and cost reduction. It is appealing to the users
of expensive polymers for exotic applications. Thus the challenge in producing different microcellular materials would lie in achieving an optimal cell
size and cell size distribution. However, performance needs to be evaluated

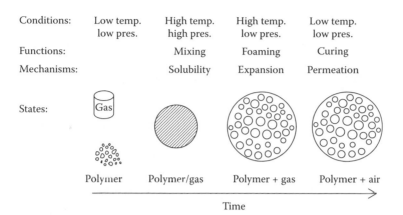

Conditions:	Low temp. low pres.	High temp. high pres.	High temp. low pres.	Low temp. low pres.
Functions:		Mixing	Foaming	Curing
Mechanisms:		Solubility	Expansion	Permeation
States:				
	Polymer	Polymer/gas	Polymer + gas	Polymer + air

Time

FIGURE 1.8
Thermoplastic foam formation path.

across a wide spectrum, ranging from very low densities (30–200 times expansion) for packaging applications, to property compliance (e.g., rigid PU foam for refrigerator doors). Conversely, a lower density accompanied by a satisfactory performance would be the logical target towards which competing foam processors should strive. Other avenues for improvements include the development of novel polymers, such as ethylene-styrene interpolymer, for example, as well as the constant upgrade of polymers; for instance, high-melt-strength (HMS) polypropylene (PP), branched polyterephthalate (PET), and metallocene-based polyethylene (mPE) all represent recent upgrade efforts.

In conclusion, it seems that the evolution of polymers, the search for blowing agents, and a better understanding of foaming will bring forth new polymeric foaming technologies in the near future, and by extension, a wider variety of foamed polymers with better environmental prospects. It is thus anticipated that the evolution of the industry for polymeric foams will bring to humankind a better and safer living.

1.6 Thermoplastic and Thermoset

Long polymeric chains can be formed by free-particle induced successive chain addition or by consecutive functional group reactions. The type and nature of chain propagation can determine the polymeric chain structure, which greatly affects its response to temperature as well as its properties. When a long-chain linear polymer is formed, the interpolymeric coiling and the van der Vaals force hold molecules together. Since both chain disentanglement and mobility increase at higher temperatures, the processing of polymers is more suitable at elevated temperatures. However, the

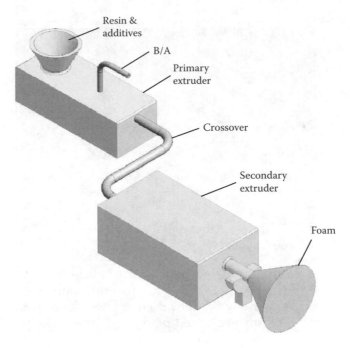

FIGURE 1.9
A typical foam extrusion diagram.

amorphous and crystalline regions can be restored when the temperature decreases. This thermally-reversible morphology is in fact a unique feature of this family of polymers, also known as "thermoplastics." At high temperatures, the fluid-like melt can dissolve a blowing agent to form a homogeneous melt solution under high pressure. After the melt is stabilized and exits into a lower pressure region, instant supersaturation induces foaming within the polymeric melt. A polymeric foam is subsequently formed upon cooling or solidification. Thermoplastics are most frequently produced via foam extrusion [13]; a typical set-up is presented in Figure 1.9, and Figure 1.10 shows how the product modulus varies with temperature.

As mentioned earlier, the functional group reaction is possible for chain length increase. A linear, planar, or even 3-D structure can be formed, depending upon the number of reactive groups and the nature of their reactivity properties. When both planar and 3-D structures are established in combination with entanglement, the resulting structure is temperature-irreversible. The dependence of the modulus/viscosity of thermoplastic foams on temperature is different from that of thermoset foams. The behaviors of both materials are illustrated in Figure 1.11. A conventional plasticator is generally not suitable for handling thermoset materials. However, if the blowing agent is of an exothermic nature, the liberated heat will not only accelerate the reaction, but will convert the product to a vapor state. The gas

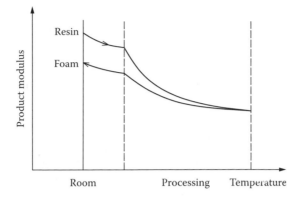

FIGURE 1.10
Product modulus variation in foam extrusion.

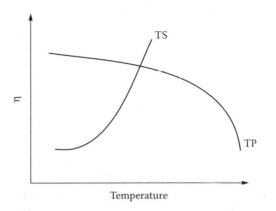

FIGURE 1.11
Viscosity change with temperature for common thermopastic and thermoset polymers.

evolution and polymerization can be independently controlled to yield desirable thermoset foaming. Polyurethane, polyisocyrunate, epoxy, and phenolics are typical examples of this foaming process.

Different thermoplastic polymers can be made into foams with inherited properties from the parent polymer. Extrusion and injection molding can be used to turn a great variety of thermoplastics into foams [34]. However, thermoset foams can be controlled in such a way that one polymer can exhibit a great variety of states and properties; such is the case with flexible and rigid PU foams. Therefore, although they serve the same market, thermoset and thermoplastic foams required relatively independent ventures into technology development. A brief comparison between thermoset and thermoplastic foams is presented in Table 1.8.

TABLE 1.8

Overall Comparison of Thermoplastic and Thermoset Foam

	Thermoplastic	Thermoset
Formula	Polymeric pellets	Reactive chemicals
Processing	Mostly continuous	Continuous and batch
Gas implementation	Gas dissolution by pressure or gas	Gas generation by
Nucleation	generation by heating	reaction
Growth	Supersaturation or pre-existed nuclei	Pre-existed nuclei
Thermal effects	Diffusion of dissolved gas	Diffusion of produced gas
Foaming time	Vaporization induced cooling	Reaction induced heating
Coalescence	Less than 3 seconds	2–5 seconds
Cell size distribution	Polymeric foam packing	Liquid foam stacking
Open cell	Normal distribution	Log-normal distribution
X-linking	Controlled by processing	Controlled by formula
Foam stabilization	None or very low	Low to high
Expansion ratio	Natural cooling	Polymerization, x-linking
Product dimension	Up to 50 times	Up to 250 times
Recycle	mm sheet to cm board	cm to m
	Reprocessable	Reuse by crushing or reduce the size by reaction

1.7 Polymeric Foam Technology

Polymers display a wide variety of properties, such as their unique thermal behaviors and their special capacity for retaining blowing agents. Throughout the second half of the twentieth century, various plasticating technologies have been developed to take better advantage of the manifold rheological benefits offered by polymers in order to convert them into useful plastic products, like thin film, plastic bags, and plastic bottles, to name a few examples. When blowing agents are involved in the production process, specially designed equipment is necessary to make polymeric foams. Three different foam formation methodologies determine the choice of manufacturing mode for creating foams: reactive foaming, soluble foaming, and quenching foaming. Table 1.9 illustrates the breakdown by percentage of the different foaming technologies currently utilized in various industries [42].

TABLE 1.9

Major Foaming Methodologies and Practices

Methodology	Principle	Practice	Market %
Reactive Foaming	Reaction/Processing	PU, X-PE, X-PVC	58
Soluble Foaming	Processing/Dissolution	PS, PE, PP	38
Melt Quenching	Solution/Quenching	Aerogel	Under 4

Each technology is characterized by particular features that enable the smooth processing of polymers and guarantee the manufacture of a stable end product.

Reactive foaming is the most popular method used in the polymeric foam industry today. According to the reactive foaming process, final foam products are yielded from a chemical reaction that initially involves a certain amount of gases, which in turn suffuses the newly polymerized matrix, ultimately creating a polymeric foam product. The chemistry is quite straightforward. For instance, a polyurethane foam contains polyols and isocyanate, and the highly reactive isocyanate can react with water to form carbon dioxide readily for expansion. In fact, variation in material formulation, such as the choice of reactants and the particular ratio in which they are combined, can result in a number of different foamed products that vary in density, product dimension, cell size, and cell nature. Each product may in fact be characterized by dramatically different properties depending on its application destination.

It is commonly known that gas can be generated via thermal decomposition. Gas generation can be adapted in the plasticating processing, during which decomposition occurs. The blowing agent that can decompose and generate gas is fed into the processor in a solid form. These types of blowing agents are known as "chemical blowing agents" (CBAs), and are widely used in the production of cross-linked polyolefin foam, rotational molding, and injection molding to make structural foams. Both polyurethane (PU) and CBA foams will be addressed at later chapters of this book.

Soluble foaming is another well-developed technology; it involves the dissolution of a physical blowing agent (PBA) in the polymeric melt. Extrusion and injection molding with PBA are typical examples of soluble foaming for thermoplastics. The main advantage of this technology is that it permits high-speed continuous processing, especially in extrusion. A specially designed screw is necessary to generate high pressure in the mixing section, which then enables the dissolution of gas in the melt and thus the formation of a homogenized melt/gas solution. This melt/gas solution is subsequently delivered to a low-pressure environment to allow foaming to occur. Another technology of note is bead (or particle) foaming technology. This technology proceeds by saturating a polymer bead with enough gas at an elevated temperature and pressure, so that later on the saturated bead is under heat again, for the encapsulated gas to expand. Sometimes, application of vacuum (or lower) pressure is necessary to assist the expansion. The bead foaming technology has sometimes been combined with mold foaming to create finished products.

Yet another technology to consider is melt (or solution) quenching. It was originated from aerogel methodology, which involves compatible liquid to expand the solid, then replacing the liquid with gas. Silicon dioxide in high temperature and high pressure to contain acid, which is then replaced by alcohol, then by air to form areogel. A typical process is presented in Figure 1.12.

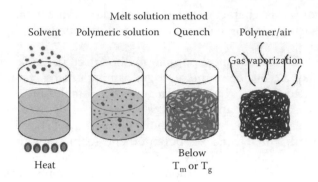

FIGURE 1.12
Melt solution method.

When a polymer is employed, the pressure and temperature requirements are not that stringent. It thus opened the door for polymeric solutions.

When the amount of the blowing agent is increased to a point of rendering the dissolution of the polymer in the blowing agent possible, which is the reverse of the previously mentioned technologies, the blowing agent dissolves in the polymer. In this case, the polymer constitutes the minor phase, and the solvent (i.e., the blowing agent) comprises the major one. The viscous solution can be quenched to a point much lower than its solidification point (or glass transition point), so that the polymer tends to solidify and form a skeleton. Phase diagram is quite necessary to understand the phase separation to determine the ratio of polymer and solvent [28]. It can be imagined the newly formed skeleton is soaked with solvent, and that it is necessary to raise the temperature and apply vacuum to drive out the solvent. An open-celled skeleton is formed in this manner. When using biodegradable or water-soluble polymers, this method may demonstrate unique medical and sanitary applications because of the unique open porous structures [43].

The above is a brief overview of all of the major foaming technologies used today. Other processes, such as mechanical blending and emulsion frothing, for instance, are examples of technologies of secondary importance. The major foaming technologies will be addressed in the later chapters of this book in great detail, which will give us a thorough understanding of the foaming process as it pertains to everything from scientific principles to pilot tests and to production technologies.

1.8   Applications and Usage

Polymer consumption has been climbing since the beginning of 20th century. In the last 25 years, the usage of foam has increased dramatically. It appears that the increase of the volumetric consumption of foams has exceeded that of polymers, even though foams are subject to tougher environmental regulations.

Foam applications have become quite diversified due to specific, yet equally dedicated, developmental initiatives. New products virtually replaced some existing products and sparkled new ideas in applications. Usually application itself find inroads for more usage. As a result, the market continues to experience new demands, some of which stem directly from consumer demands. The foam application started in floatation, then packaging materials were developed with foams around the 1970s. After this, many applications were developed in the areas of construction, automotive, sports, electronics, medical devices, and so on. Lately it was found that microcellular foam could not only enhance the property/weight ratio of the foam, but that it might also contribute to the redefinition of thermoplastic material structures in general. If the cell size is decreased from micro to nano, this could indeed have a tremendous impact on polymeric materials. In other words, the nanocellular foam with unique properties could further expand the application spectrum into specialty and material domains.

Into the 21st century, developed countries continue to demand high standards of living, which creates opportunities for new products (e.g., recreational, automotive, medical, sports and so on) to enter the market. Polymeric foams are among the first choice to meet this need because of their unique properties. On the other hand, developing countries will also catch up this trend of the new product developments. This development of new products may become a major driving force behind the global consumption of foam. Packaging, electronic, and construction were good examples of this trend in the past, although it is expected that many other applications will use more plastic foams to meet various demands of society.

It is fair to say that polymeric foams are not absolutely necessary to sustain life, but they become indispensable to enrich our living. When demands and consumption continue to sour, let us not forget that technology, business, and regulation should work together to ensure a balanced and healthy growth of the foam industry.

2

Thermodynamics and Kinetics

Foaming polymers involves materials in various phases, states, and changes between states and phases. In general, the changes are from stable state to unstable state, then to another stable state. Thermodynamics and kinetics become two fundamental principles governing the changes. The former defines the ultimate system conditions we can accomplish, and the latter shows the rate process to reach the conditions. Basically, they go hand in hand. It is of vital importance to understand thermodynamics and kinetics, before we investigate engineering and processing parameters.

It is known that gas and polymer are quite different in nature as well as in thermodynamic responses. Gas basically follows temperature-pressure-volume gas law:

$$P V = n R T \qquad (2.1)$$

where, n and R denote mole and gas constant, respectively. The equation has been widely used to describe the state of the gas, and often as a useful reference equation for foaming volume calculation. On the other hand, the volume of polymers or polymeric melt is very inert to pressure variation. When gas and polymers are together, drastic compromise can be imagined in forming a system.

This chapter begins with equilibrium, then non-equilibrium, which naturally involves kinetics, transport, and rheology. In each state, certain parameters become the governing parameters, and proper attention to each will help us comprehend the intriguing polymeric foaming without getting lost in the variety of competing mechanisms.

2.1 Equilibrium

Foaming in polymers is a thermodynamics-driven kinetic phenomenon or process. It is both necessary and important to understand the under-ruling thermodynamics and its parameters. Before we proceed, a brief definition

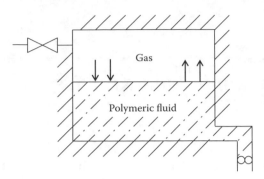

FIGURE 2.1
Equilibrium state in a closed gas/melt system.

of state and system is requisite. State is inherent to material regardless of phases with given properties under given conditions. A system is defined by surrounding the material(s) with a boundary, which facilitates investigation [44]. A simple polymer/gas system is presented in Figure 2.1.

A polymer consists of a multitude of monomers, and the long-chain polymer, instead of straight line alignment, often coils. When a group of polymers stick together to form polymeric material, micro-spaces out of coiling and chain-ends exist, which are referred to as free volume or holes. When gas molecules come into contact with polymer molecules in a liquid or solid phase, they tend to secure residing in the free volume. Initially, gas molecules fill in the holes, which is known as the Langmuir mode [45]. As these holes become saturated with gas molecules, the gas molecules start to swell the polymeric matrix, during which the gas incoming rate slows down. After a sufficiently long interaction, the incoming and outgoing gas rates become equal, which state is referred to as equilibrium.

Equilibrium is a function of the surrounding temperature and pressure, which govern the activity of materials. Conversely, the ratio of vapor pressure and solubility for a single vapor dissolving in a melt phase at a given temperature and pressure is referred to as the equilibrium constant (E_c), which can be expressed as follows:

$$E_c = \text{(gas concentration in vapor)}/\text{(gas concentration in liquid)} \quad (2.2)$$

The equilibrium constant defines what the ultimate stable state (i.e., the maximum dissolution or separation) is, which is a very useful thermodynamic parameter. It can be used to calculate, the amount of gas dissolution at the system pressure, and the degree of super-saturation (oversaturation) for bubble nucleation and growth, which is a primary driving force for foam kinetics.

Based on statistical thermodynamics, Flory and Huggins developed a theory to describe the vapor pressure of a polymer melt/gas system, especially for low concentration of gas, giving rise to the following relationship [46,47]:

$$\ln(P_g/P_g^\circ) = \ln\phi_g + 1 + \chi \tag{2.3}$$

where, P_g°, ϕ_g, and χ represent the vapor pressure of the pure gas, the gas volume fraction, and the interaction parameter, respectively. Let's now consider the Henry law constant (a well-known gas/liquid constant), which is given by:

$$K_w = P_g/W_g \tag{2.4}$$

where P_g and W_g denote the gas partial pressure and gas weight fraction in the liquid, respectively. Hence, by combining Equations 2.3 and 2.4, the Henry law constant can be expressed as [48]:

$$K_w = P_g^\circ(\rho_m/\rho_g) \exp(1+\chi) \tag{2.5}$$

where ρ_m and ρ_g denote the density of polymeric melt and of gas, respectively. Since the magnitude of the density ratio ranges between two and three, it follows from Equation 2.5 that the vapor pressure and the interaction factor have the greatest influence on the equilibrium constant. Consequently, simple but precise experiments can be conducted to determine the equilibrium constant for particular gases of interest with sufficient accuracy as shown in Figure 2.2 [49–52]. This chart is extremely useful for practice. When solubility at two different temperatures is known, the linear mode can help us get the solubility for other temperatures through intrapolation or extrapolation. Equilibrium pressure can be calculated to serve as a good reference

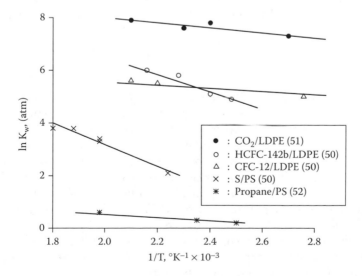

FIGURE 2.2
Variation of Henry's law constant vs. inverse absolute temperature (data collected from [50] to [52]).

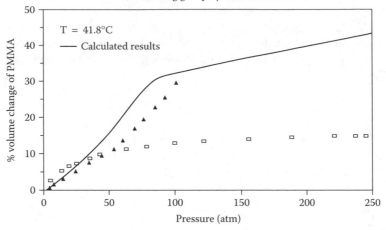

FIGURE 2.3
Polymer volume increase out of gas dissolution at different pressures. (M.B. Kiszka et al., Modeling High-Pressure Gas-Polymer Mixtures Using the Sanchez-Lacome Equation of State, *J. App. Poly. Sci.*, 36, 1988. With permission.)

for equipment design. More common solubility data can be found in the literature [53,54].

Sanchez-Lacomb developed a state equation to correlate pressure, volume, and temperature, which can be applied to gas in polymeric liquid. It assumes holes and lattice to generate material parameters. At equilibrium the chemical potential is at a minimum, and the equation of state becomes [55]:

$$(1/v_r)^2 + P_r + T_r[\ln(1-1/v_r) + (1-1/r)/v_r] = 0 \qquad (2.6)$$

where v_r, P_r, and T_r are reduced volume, pressure, and temperature; r denotes sites occupied by a molecule in the lattice. Figure 2.3 shows volume change of polymeric melt at different pressures [56,57]. It can reach over 30% volume increase at 100 atm, which is not unusual for extrusion and injection molding. Gas dissolution tends to plasticize polymeric melt, which reduces the apparent viscosity to decrease flow pressure for a given volume rate. But gas dissolution evidently enhances apparent volume. That means at a given flow geometry, the pressure naturally increases. When the latter outperforms the former, it develops a lower viscous system, and yet a higher processing pressure under the same processing volume. As a result, processing pressure increases as gas content increases in the melt. A typical example is the gas/melt in the shaping region of the extrusion process.

2.2 Super-Saturation

When surrounding conditions change, the state of the system changes accordingly. As pointed out in Chapter 1, foaming is a dynamic phenomenon that dissipates disturbances in order to restore a low-energy stable state. Hence, the state of super-saturation or under-saturation can be regarded as a thermodynamic parameter to characterize the change of conditions in the surrounding environment. Imagining a system is saturated after long enough time, when the temperature increases or the pressure decreases, the already saturated system becomes supersaturated, which has a natural tendency to adjust itself to resume a stable state.

It is important to note here that a supersaturated state is a necessary condition for bubble formation. In this context, it is useful to define the degree of superheat (SH) as follows:

$$SH = K_w W_g - P \tag{2.7}$$

where the new pressure is denominated by P. After long enough time, SH becomes zero, which means equilibrium is established. Any change of surrounding conditions can provoke change of SH. Thus, for example, when vacuum is applied, P becomes lower, thereby providing adequate conditions for super-saturation to take place.

$$SH > 0 \tag{2.8}$$

However, if P remains unchanged, the increase of temperature will result in a higher K_w, which will also result in a positive SH. When SH is greater than zero, the system itself is inclined to adjust to zero through diffusion, vaporization, foaming, etc.

In general, the degree of superheat should be regarded as a primary driving force for bubble nucleation and growth [58]. Its magnitude determines both the bubble nucleation and growth rates. In nucleation, it follows a diffusion induced threshold which turns into an exponential expression to be elaborated at a later section. The matter in fact is at a low or modest superheat, the corresponding nucleation rate is very low or too low to be detected. Mathematically, modest exponent is necessary to make noticeable nucleation rate. In fact, at a low SH, foaming may become non-observable. In other words, a modest SH may not be sufficient, and other energy inputs are necessary to facilitate nucleation. Nonetheless, a critical superheat appears to be necessary to drive new phase into existence via vigorous nucleation.

Thermoplastic polymers are known for their unique temperature-dependency of rheology, which is often thermal reversible. Common plasticators, generally screw driven machinery, cannot only melt the polymers, but are capable of building up pressure, which is necessary in pumping the viscous polymeric melt and in dissolving gas and keeping the gas in the melt. The inherent thermoplastic properties and high pressure capability in the plasticator make a unique combination extremely important for foam processing.

This structure dependency on temperature makes them suitable for processing during which the blowing agent can be introduced and kept with the polymer at a higher pressure and temperature to establish a homogeneous state. When pressure is released, the critical superheat is virtually met, and swarms of bubble occur. For instance, in extrusion, at the point where the extruded strand meets the atmosphere at the end of die, the sharp reduction in pressure renders a significant positive superheat for foam formation.

This is also true for static foaming, i.e., when a gas-saturated polymer is moved to a thermal bath to create a heat-induced super-saturation to form bubbles [59]. However, unlike the instant-pressure-reduction-induced foaming concept, the heat-induction concept takes a substantially longer amount of time owing to the poor heat conduction nature for the polymer. Consequently, the pressure induced superheat is currently the dominant foaming mechanism used for the manufacture of thermoplastic foams.

It is noted that in many industrial foaming processes superheat is a process variable, since either pressure or temperature, or both are changing in the process, which makes the nucleation agreement between theoretical calculation and actual results very difficult to obtain. At any event, the superheat concept, overall superheat and critical superheat, overrides in most foaming processes.

2.3   Reactions and their Kinetics

To create foamed products, especially polymeric foams, at least two conditions—processing and material—should be met. Firstly, a mechanism for even gas-phase generation and/or implementation into a denser medium should be put in place, and secondly, the material intended for foaming should be such to hold the gaseous bubbles dispersed within its matrix. It is a staged process; gas molecules in the polymer first, then gaseous bubbles in the polymer.

Gas-phase formation within a medium is a well-known foaming phenomenon, however, as discussed before, there are a variety of methods to implement the gas in the polymer. But usually it is not sufficient to make polymeric foams. It is important to emphasize here that only a material with an adequate material strength can yield quality foamed products. Otherwise, the

TABLE 2.1

Various Types of Reactive Foaming Process

Type	Mechanism	Application
Decomposition in polymeric melt	Heat-induced reaction High pressure foaming Natural Cooling	Extrusion with CBA Injection molding
Decomposition in polymer	Heat-induced x-linking Heat-induced gasification Low pressure foaming Forced Cooling	X-linked PE foam
Reaction and polymerization	Chemical reaction Long chain formation	Thermo-set PU foam Polyisocyanurate Foam

foaming could exceed the material strength to cause bubble rupture or cell wall weakening so that the polymer cannot sustain the expanded dimension. It could result in obtaining a transient foaming phenomenon until vaporization and/or bubbling is completed. Anyway, common reactive foaming methodologies for making polymeric foams are listed in Table 2.1.

This section focuses on the gas-phase generation through the reaction within the polymeric material. Gas phase exists almost everywhere. To generate a gas phase, such as vaporization, decomposition, reaction, entrainment, etc., is not an issue. However, generating gas phase evenly within a liquid requires attention and skills. Reaction and decomposition are common concepts in gas evolution within a liquid phase. A typical example is the decomposition of sodium bicarbonate in a solution:

$$2NaHCO_3 \rightarrow Na_2CO_3 + H_2O + CO_2 \qquad (2.9)$$

When the reaction occurs in a concentrated liquid, and the temperature is above the boiling point of water, both water and carbon dioxide convert into a gaseous phase within the medium. The foaming has thus occurred. It should be pointed out that the mixing time is a critical parameter for the second or higher degree reaction to allow reactants to contact and to interact long enough for the reaction to complete. A common chemical engineering practice in reactor design is to consider the ratio of the reaction time and the reactor residence time as a guideline in reactor design, size and geometry [60].

Since polymeric melts are inherently viscous, the first-degree decomposition reaction is more desirable and feasible within the polymer processing equipment. When the evolved gas phase is above the solubility limit, a vapor phase comes into existence. As has been mentioned earlier, the solubility is strongly dependent on the surrounding pressure. Thus, a processing facility characterized with both high temperature and high pressure appears to be desirable for accommodating the decomposition reaction within the pressure-reduction-induced foaming concept. A good example is the chemical blowing agent (CBA)-based foaming process in extrusion and injection molding [61–63]. CBA is capable of releasing simple chemicals at a higher temperature via decomposition. Under high temperature, the chemicals virtually become blowing agents.

It is known that the reaction is endo- and exthothermic in nature, depending on heat absorption or evolution throughout the reaction. In general, the evolution of carbon dioxide is common during the decomposition of endothermic CBAs, whereas a common product of the decomposition of exothermic CBAs is nitrogen.

A common extho-thermic CBA is azodicarbonamide, which decomposes around 210°C. Its main decomposition reactions are:

$$2H_2NCONNCONH_2 \rightarrow H_2NCONHNHCONH_2 + N_2 + 2HNCO$$

$$(2.10)$$

$$H_2NCONNCONH_2 + 2HNCO \rightarrow H_2NCNOHNHCONH_2 + N_2 + 2CO$$

$$(2.11)$$

The majority gas evolutions are nitrogen and carbon monoxide. Under standard temperature and pressure (STP), it generates 220 ml gas per gram of azodicarbonamide, which is widely used in polyurethane, polystyrene, and X-PE foam industry.

At high volume expansion foaming and/or when using a heat-sensitive polymer, an endothermic reaction is required. A recent development involves a synergistic system including a blend of citric acid and sodium bicarbonate, which reaction is shown below:

$$C_6H_8O_7 + 3NaHCO_3 \rightarrow (C_6H_5Na_3O_7)\ 2H_2O + 3CO_2 + H_2O \qquad (2.12)$$

The reaction begins at 160°C and is completed at about 210°C. The weight ratio between citric acid and sodium bicarbonate is around 44:56. One gram of the blend can generate about 120 cm^3 blowing gas [61, 64]. That means 1% of the above can generate approximately two times volume expansion, or 50% void.

Since some common inorganic blowing agents, such as nitrogen and carbon dioxide, possess a much greater volatility, aggressive foaming is thus anticipated, which makes polymer strength more of a challenge. Chain extension or chain x-linking via irradiation treatment or free particle initiation became a good practice to enhance polymer strength in order to withhold the decomposed gas. The requirement for successful foaming is to induce cross-linking of the polymer prior to the decomposition of the CBA. In this context, peroxide is a particularly good candidate for a free particle generator because it decomposes at a lower temperature than most CBAs. A two-stage heating profile was developed to produce X-PE foam by Sekesui Plastics

TABLE 2.2

Selected Properties of Common Chemical Blowing Agents

Chemical Composition	Decomposition Temp. °C	Main Evolved Gas
Azodicarbonamide	204–212	N_2
Citric Acid/Sodium Bicarbonate	170–210	CO_2
Benzene Sulfonyl Hydrazide	157–160	N_2
5-Phenyltetrazole	240–250	N_2
Sodium Borohydride	*	H_2
*: activated by immersing in water		

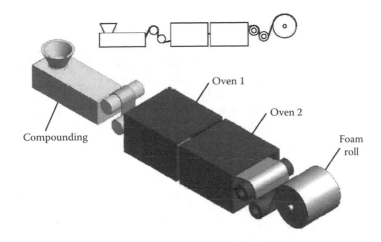

FIGURE 2.4

A simple schematic of chemically x-linked PE foam process; side and angle views.

since the 1970s [65]. A brief summary of common CBA kinds and decomposition temperature ranges is listed in Table 2.2.

As depicted in Figure 2.4, it is possible to establish a continuous two-stage heating process for cross-linked foams. Polyethylene has a stable free particle to allow chain cross-linking to dominate over chain scission, which becomes a good example in making cross-linked PE foams out of a sheet/oven process.

Another well-known technology is reactive foaming via polycondensation to evolve gaseous by-products. During this process, polymer chain lengths continue to grow. Since the amount of gas and the degree of polymerization fit nicely to the foaming and foamed plastic requirements, the exothermic reaction promotes gas volatility for further expansion while the material strength permits. In this case, heat becomes a favorable factor in foaming. For this reason, reaction followed by chain extension to build up material strength became a very popular methodology in producing thermosetting foams. Polyurethane foams are a typical example. The suggested reaction route is:

$$R-N=C=O + H-O-H \rightarrow RNHCOOH + R-NH_2 + CO_2 + heat \quad (2.13)$$

$$R-N=C=O + R'-CH_2-OH \rightarrow R-NH-COO-CH_2-R' \quad (2.14)$$

$$R-N=C=O + R-NH_2 \rightarrow R-NH-CO-NH-R \quad (2.15)$$

Importantly, since the approximate heat release per mole of water is 47 kcal, it is capable of heating up the mixture above the boiling point of water to generate extra foaming. The exothermic urethane reaction makes itself unique in using water expansion for additional expansion of thermoset. Consequently, PU foams for packaging application are capable of expanding above 200 times, while they can also be controlled in terms of cell size, structure, and strength for prescribed applications. The foaming mechanisms suitable to cross-linked PE foams and PU foams are listed in Table 2.3.

Although under the same chemistry, the number of the functional group, the length of polyol, and the use of catalysts can result in a wide variety of foamed products in terms of density and property. After 70 years' practice, PU foam nowadays has become the largest section in polymeric foam industry. A simple concept for making flexible, semi-rigid, and rigid PU foam and their density ranges is presented in Table 2.4.

2.4   Melt/Gas Rheology

Gas and polymer are completely different materials, yet under high pressure and high temperature, polymeric melt is capable of taking certain gas phase

TABLE 2.3

Reactive Foaming Principles and Processes

Type	Foaming Path	Methods	Reactor
X-linked PE Foam	X-linking+Gas Generation	Heating	Oven
PU Foam	Gas Generation+Chain Buildup	Mixing	Mixer+Mold

TABLE 2.4

Simple Comparison between Flexible and Rigid PU Foam

Characteristic	Flexible PU Foam	Rigid PU Foam
Polyol Molecular Weight	1,000 to 6,500	400 to 1,200
Polyol Functionality	2.0 to 3.0	3.0 to 8.0
Isocyanate	TDI based	MDI based
Density Range, kg/m³	5 to 200	20–800
Key Property	Absorption	Insulation

to form a homogeneous solution for foaming. At any event, the presence of gas in the melt is a critical issue in processing thermoplastic foams. First of all, it is very dynamic itself and, in addition, under motion. The viscous nature of polymeric melt can easily generate shear-induced heat to cause flow instability (or gas phase separation) concerns. This is a minor concern in foaming of thermoset because it, in general, involves less flow and more foaming in comparison with the thermoplastic foaming. When gas is dissolved in the polymeric melt in a closed system, it causes the polymeric structure to swell, which, in turn, alters the melting and flow characteristics. The effects of hydraulic pressure exerted outside the polymer on the structure of polymer will be much less than those of the dissolved gas when enough gas is dissolved into the polymer. As a result of the presence of dissolved gas in the molten polymer, the polymer crystallization is depressed, and the flow activation energy is reduced so that it decreases flow onset while increasing the flow rate.

Gases have a different response from polymeric melts toward outside disturbance. When gas molecules disperse within the polymeric structure to form "composite units," their presence can certainly alter the overall capability in dealing with internal and external changes. In melt flow, polymeric clusters change their shape and position to respond toward the external pressure or internal momentum transfer. Free volume theories have been established to describe its temperature dependency of viscosity, especially useful when close to the glass transition temperature, T_g. However, in a molten state (above the T_m), the space generated by chain moving definitely exceeds the free volume, which may not account for the flow behavior. But the presence of gas evidently enhances the overall activity of the polymer/gas system, which corresponds to a higher system pressure.

It is reasonable to find that gas/melt mixtures possess loose chains folding and aligning over each other. As a result, reduction of shear viscosity appears with respect to addition of blowing agent. As illustrated in Figure 2.5 [66], the reduction of viscosity increases at increased amount of gas molecules in the polymer regardless of its amorphous or semi-crystalline structure. It is likely that gas molecules disperse in the amorphous region, which can be envisioned as particle-hole or hole-lattice. Flory-Huggins, Sanchez-Lacombe, and Simha-Somcynsky theories and equations were thus developed as mentioned in Section 2.1 [67]. These theories were originally intended to describe pressure-volume-temperature behaviors, and lately, these were attempted to be implemented for describing the shear viscosity variation due to gas dissolution [68-69].

Clearly, viscosity reduction facilitates processing in facilitating flow and in reducing viscous heat generation. But, on the other hand, the processing pressure should be kept higher than the vapor pressure corresponding to the dissolution. Otherwise, phase separation can occur and thereby cause flow stability concerns [70], which can easily cause premature phase separation rather than decent foaming.

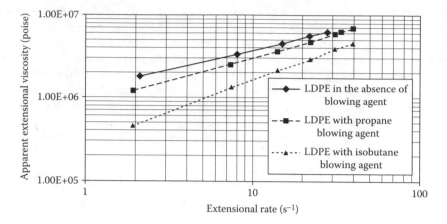

FIGURE 2.5

Extensional viscosity reduction for polyethylene with various blowing agents at 150°C. (Adapted from [71].)

The extensional viscosity has a similar reduction at adding the blowing agent as seen in Figure 2.6 [71]. As a result, the gas/melt solution essentially illustrates a reduced melt strength, which adversely affects the foaming capability. Therefore, it becomes a necessary practice to lower the processing temperature profile and thereby to increase the material strength to a point that foaming can be attained.

In brief, the addition of gas causes structural modifications of the polymeric melt, which leads to polymer rheology variations that may affect both processing and foaming in different ways. Plasticization (viscosity reduction) accompanied by swelling (specific volume increase), expansion and latent heat in vaporization at the same time, are good examples. A brief summary in gas dissolution is presented in Table 2.5. Therefore, caution has to be exercised to ensure smooth processing and successful foaming.

Another related subject is the depressed glass transition temperature at increased blowing agent. It is conceivable that free volume may be enlarged out of the active blowing agent indwelling, and its effects become pronounced as its level increases. By the same token, when the blowing agent is introduced in the molten state, its presence evidently slows down the formation of crystalline and/or enhances chain ends movement. As a result, more heat removal is needed to solidify the polymeric fabric. Figure 2.7 shows the crystallization temperature shift for LDPE [72]. It is not surprising in seeing the slight increase in melt peak after relax and reheat of foam sample.

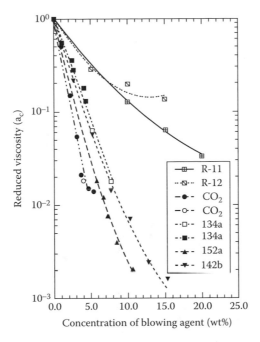

FIGURE 2.6
Shear viscosity reduction for polystyrene with various blowing agents at 150°C (Adapted from [66]).

TABLE 2.5

Effects of Gas Dissolution in Polymer and Its Processing

Material Property:
  Viscosity Reduction, T_m and T_g Depression, Specific Volume Increase,
  Diffusion Coefficient Increase, Inherent Vapor Pressure

Processing Parameter:
| Homogenization: | Adequate Mixing Zone |
| Stabilization: | Heat Removal, Uniform Thermal Distribution |

2.5 Crystallization Kinetics

In a continuous plasticating process, polymer changes from solid to molten state, and ultimately to solid state for applications. The former is called melting, and the latter solidification or crystallization. In general, melting does not involve gaseous blowing agent, but foaming and solidification occur at the same time with drastically different rates. Although foaming is usually an instant process and solidification relatively slow, the material

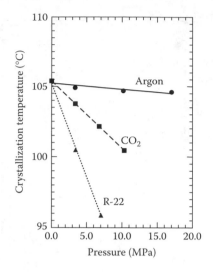

FIGURE 2.7
Depression of crystallization temperature of LDPE at different pressures for
Argon, CO_2, and HCFC-22 [72]. (S.K. Dey et al., Effect of Physical Blowing Agents on Crystal-
lization Temperature of Polymer Melts, *Annu. Tech. (ANTEC) Conf. Proc. Soc. Plas. Eng.*, 1994.
With permission.)

strength sets the boundary for foaming activity for a useful foamed product.
That means from vigorous foaming phenomena to foamed product, crystallization or solidification plays a critical role in degree of expansion and in
integrity of cellular structure.

When molten polymer begins to cool, chain movement slows down to a
point spherulite starts to expand to form crystalline or amorphous structure.
In general, crystalline is formed within a narrow temperature window, and
amorphous much wider, during which foaming begins and progresses.
Structural formation provides a boundary to foaming to encapsulate the
gaseous voids.

Depending upon the polymer chain structure, crystalline can be formed
within a limited period to cause resistance for bubble expansion to secure
fine cell structure, or aggressive foaming tends to excel the yield point of
the crystalline to disrupt expansion for a rough structure. The competing
mechanisms between expansion and material strength to hold expansion
become an interesting kinetic subject for optimal foaming and fine structure.

The isothermal solidification rate can be expressed by the Avrami equation
[73,74]:

$$x(t) = 1 - \exp(-kt)^n \qquad (2.16)$$

where $x(t)$ is the fraction of solidified material in spherulitic state at time t,
and k and n represent nucleation parameter and geometry integer, respectively. k includes transport and nucleation activation energy barriers, which

is a strong function of temperature. Oftentimes, processing profile can dramatically vary the solidification rate. The solidified fraction directly affects material strength, which governs the bubble expansion. If it solidifies too fast, limited expansion is expected; if too slow, then over-expansion. Proper temperature control on account of blowing agent amount and volatility is crucial in commercial foaming process.

Keller argued in his phase tranformation thesis [75], that the ultimate thermodynamically stable state is hardly ever achieved for polymeric solidification. Basically we are dealing with metastable states, which metastability manifests itself in phase transformation in either thermodynamically stable phase or early non-stable phase. That means it could be the transient state or the kinetically preferred state. With these chain folding and spherulite growth in crystallization can be better understood as indicated in Figure 2.8. Both have profound influence for the phase transformation, and, in turn, affecting morphology and property of the polymer.

Also suggested was that crystallinity could be adopted as an effective way to enhance melt strength, especially when properly controlled in blend formula and rate of temperature decrease. Yamaguchi pointed out the controlled conditions for quantitative crystallinity are difficult to set up, and some is forthcoming from academia [76]. This simply offers direct benefits for semi-crystalline polyolefin foaming.

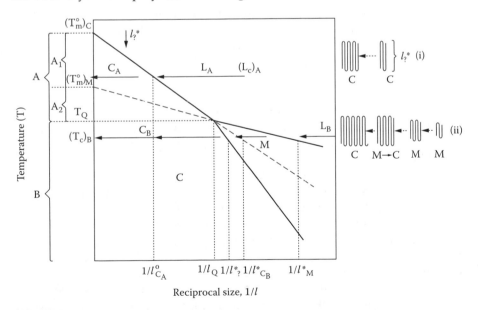

FIGURE 2.8
Isothermal phase stability diagram; crystal growth in temperature, T, vs. reciprocal crystal size, $1/l$. (—) stable-phase demarcation lines, (-----) metastable-phase demarcation lines, (→) showing isothermal growth pathways. Two pathways are illustrated; one above and one below the triple point T_Q; which are representative of the growth regimes A and B, respectively. M denotes metastable and C crystalline. (A. Keller, An Approach to Phase Behavior, *Macromol. Symp.*, 98, 1995. With permission.)

2.6   Transport Phenomena: Diffusion and Permeation

Diffusion and permeation are relevant phenomena in processing, foaming and post-foaming. In fact, diffusion is the primary transport mechanism in bubble growth regardless of thermoplastic or thermoset foaming. Driven by the concentration difference across the melt/bubble interface, diffusion occurs at all times. As the bubble starts to grow, the increase in the volume of bubble induces a gas concentration decrease in the bubble to further encourage diffusion of gas from the polymer matrix to the gas bubble. "Chained" reaction appears in the growth dynamics. It is noted that volatility primarily accounts for expansion, and diffusion for mass transfer. We may have a fast expansion, and yet slow diffusion, or slow expansion and fast diffusion, depending on the volatility and diffusivity combinations.

The diffusivity (or diffusion coefficient) follows the Arrhenius thermal-dependency as shown in Equation 2.10:

$$D = D_0 \exp\left(-\frac{\Delta E}{RT}\right) \qquad\qquad (2.17)$$

It is thus not surprising to find that as the polymeric melt cools down into a solid, the gas diffusivity decreases for several orders of magnitude (i.e., 10^{-5} cm^2/s to 10^{-9} cm^2/s). The diffusion characteristic time is defined as:

$$\lambda_D = l^2/D \qquad\qquad (2.18)$$

where l represents diffusion path. The diffusion characteristic time is a simple, but useful parameter to characterize the diffusion kinetics. Two areas are particularly affected. One is the mixing in the plasticator, and the other is bubble growth.

When a blowing agent dissolves in the polymer, it is necessary to calculate how much space is needed to achieve homogenization. In foam injection molding and extrusion, when PBA is introduced through the pumping unit, the gas stream is forced to contact the polymeric melt. Screw revolution can break up the injected gas streams into gas pockets, the size of which determines how much time for dissolution is necessary [77,78], so that the size of the mixing zone can be determined.

As the blowing agent dissolves in the polymer, the diffusion coefficient becomes much higher owing to the enhanced diffusion path in the presence of the gas in the polymer. During the cell growth, the concentration of gas in the polymer matrix decreases, and this results in a reduction of the diffusion coefficient. As illustrated in Figure 2.9, the reduction of diffusion coefficient could be as much as one order of magnitude [79]. Furthermore, the foam gets cooled during cell growth. Therefore, diffusion could be much

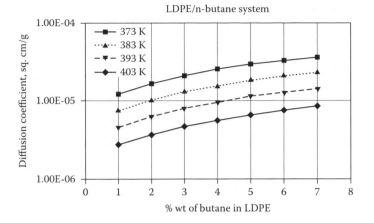

FIGURE 2.9
Effect of butane concentration on the diffusion coefficient in LDPE at different temperatures [79].

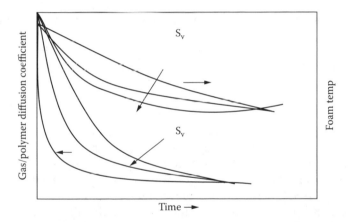

FIGURE 2.10
Diffusion coefficient and temperature variation during aging of freshly made foam; S_v represents surface to volume ratio. In general, diffusion decreases from w^{-5} cm²/sec (gas/melt) to 10^{-9} cm²/sec (gas/solid).

slower in the later stage of growth. Considering concentration and temperature effects, the diffusion parameter can be depicted in Figure 2.10.

When the foam is made, the cells are filled with the blowing agent. After solidification, the blowing agent will diffuse out and the air will diffuse into the cells. Mutual diffusion of air and blowing agent continues until the blowing agent is completely replaced by the air. The concentration gradient of blowing agent determines the diffusion rate, which decreases as expansion continues. However, solubility of gas in polymer plays an important role, which basically determines the amount of gas in the polymer. In fact, the solubility of gas in the solid affects the availability of penetration channel. It is known that, unlike diffusivity, solubility increases as temperature

TABLE 2.6

Permeability of Selected Polymer/Gas Systems (collected from [12, 81–83])

System		Permeability, $10^{10} \times cm^3 (STP) cm/cm^2$ sec cmHg
Polystyrene/	$CHClF_2$	0.17 [12]
	N_2	0.79 [81]
	CO_2	10.5 [81]
Low Density Polyethylene/	$CHClF_2$	5.67 [12]
	C_2H_6	6.81 [81]
	C_3H_8	9.43 [81]
	$n\text{-}C_4H_{10}$	45.4 [83]
Polypropylene/	$CHClF_2$	0.0025 [12]
Polyurethane/	$CFCl_3$	0.44 [82]
	N_2	3.4 [82]

decreases. It can be imagined that permeability requires the combination of the speed and the available path.

The permeability is defined as the product of the solubility (S), and the diffusion coefficient.

$$Pm = S \times D \qquad (2.19)$$

The permeability can be perceived as the product of the available channel and transfer speed. It becomes a critical parameter for foaming when the blowing-agent safety, environment, and property concerns are imminent [80]. The permeabilities for common polymer/gas systems are presented in Table 2.6 [12, 81–83]. If the permeability is too high, it may cause a dimensional stability concern because of the resultant shrinkage. But if it is too low, it causes safety concerns in case the blowing agent is flammable in nature. It typically takes months or years for polyethylene foams to restore the dimension [83,84]. Ideally, a blowing agent should have a permeability equal to that of air, and it should also be free from the flammability and environmental concerns.

Additives are sometimes necessary for changing the diffusion rate. Fatty acids and esters tend to form a coating on the cell wall to slow down cell-to-cell permeation, when air permeation is less affected. It becomes a common practice in polyethylene foam extrusion [85,86].

3

Foaming Fundamentals

3.1 Physical Blowing Agents

Polymeric foam is generally characterized by blowing agent indwelling and expansion within the polymeric matrix. Unstable foaming like boiling occurs and must be sustained by the surrounding polymeric material to form a stable cellular product. In most cases, blowing agent is virtually indispensable in the polymeric foaming process. There are a great variety of organic and inorganic blowing agents suitable for the process. From the nature of gas formation, it can be classified as physical blowing agent (PBA) and chemical blowing agent (CBA). The former is generally referred to as a variation of state, such as saturated liquid to liquid/vapor, vapor, then supercritical fluid state, in the processing of the blowing agent. The blowing agent never changes its composition, except the state. However, CBA is known by its formation path, such as heat-induced chemical decomposition. It begins in solid state, then evolves into gas when heat-activated decomposition occurs. In general, inorganic volatile gases, nitrogen and carbon dioxide, are evolved as the main components. In short, both PBA and CBA have been well established for specific foaming processes.

PBA is known with its suitability in the foaming process and foamed product. In the early days, it was credited by its superior solubility in thermoplastic polymers. Munters and Tandberg [18] disclosed a blending method and its immediate implementation in the foaming of polystyrene as early as 1935. Almost at the same time, halogenated hydrocarbon was recognized as a very friendly and stable agent for cryogenic system and, in turn, polymeric foaming. Since then, foaming with physical blowing agents became an intriguing development subject. During World War II, quite a few floating devices were made of PBA for military usages. In the 1950s, the foundations for foam extrusion and polyurethane foam with PBAs as auxiliary blowing agents were firmly laid.

PBA played a critical role in the very beginning of the development of polymeric foams, especially when the plasticating equipment was in its early stage. Therefore, the contributions of the halogenated blowing agents should

TABLE 3.1

Considerations in Selecting
Physical Blowing Agent

1) Physical
   Volatility (Boiling Point)
   Critical temp.
   Latent heat
   Conductivity
2) Chemical
   Henry's law constant
   Reactivity
   Stability
3) Transport
   Diffusivity
   Permeability
   Plasticization
4) Safety
   Flammability
   Toxicity
5) Environmental and Regulatory
   Ozone
   Green House
   FDA
   Smog
   Odor
6) Availability
   Cost
   Easy to handle
   Storage

be recognized, although, in the 1980s, most halogenated hydrocarbons were found detrimental to the upper atmosphere ozone distribution. Without these blowing agents, today's foam industry would have had a much narrower application spectrum.

According to current beliefs, a successful physical blowing agent must at least possess the characteristics listed in Table 3.1, and the list will continue to be revised as new foaming technologies and new regulations on account of their emission and environmental impact are developed. Some requirements will be relaxed as processing technology continues to improve, and some others will be tightened when new findings justify it. In general, physical blowing agents are volatile organic chemicals (VOC) including hydrocarbons (HC), chlorofluorocarbons (CFC), hydrochlorofluorocarbons (HCFC), and hydrofluorocarbons (HFC). Inorganic gases, such as carbon dioxide, nitrogen, and argon, have been tempted as solo blowing agents and blends with other VOCs.

Physical blowing agents (PBAs) can be incorporated within the polymer matrix using various methods: (i) physical blending and physical dissolution, (ii) physical blending and chemical decomposition, (iii) physical dissolution, and (iv) chemical reaction and encapsulation. Among these, physical blending and dissolution is considered the most commonly implemented method in the industry of polymeric foams. Under high pressure, and sometimes elevated temperature, a physical blowing agent can be compressed as a critical or super critical fluid, depending on the processing temperature and the critical temperature of the fluid. It then contacts and dissolves into the polymeric melt to form a saturated polymer/gas system, which can foam when subjected to a lower pressure (or higher temperature) environment. Polystyrene and polyolefin foam are good examples.

Most commonly, a hydrocarbon blowing agent is injected into the extruder to meet the molten polymer under a high pressure and high temperature. After thorough mixing, a homogeneous solution is achieved. When the solution is forced through an orifice to outside atmosphere, a sudden pressure reduction occurs, which automatically builds up a high supersaturation to convert the dispersed gas molecules into gas bubbles. Fast expansion and slow cooling are characteristic of the foam extrusion and foam injection molding. Namely, as soon as the polymer is cooled and sufficiently solidified to build up strength to hold the bubbles, a foam product is made. Due to the saturation of blowing agent in the fresh foam, a counter-diffusion with the surrounding air naturally occurs. As a result, blowing agent concentration in the cell continues to decrease as opposed to the increase of air concentration. Eventually, i.e., after a sufficiently long aging time, the foam will consist solely of air voids dispersed throughout the polymer matrix.

A chemical reaction can also favor PBA foam expansion. In exothermic reactions, the evolved heat can certainly enhance the volatility of the existing PBA and thereby promote expansion. For example, in polyurethane foaming, the initially formed volatile gas can cause much higher expansion due to the heat-enhanced volatility and diffusivity. CFC-11 or other PBAs were usually added as auxiliary blowing agents to enhance PU foam expansion. Its high solubility in PU combined with the fact of fine cell structure makes a slow permeation of blowing agent. The very favorable high thermal resistance of CFC-11 made PU/CFC-11 foam excellent insulation characteristics for applications in construction and appliance industries. Since CFC-11 encountered ozone depletion issues, it has been a great challenge to find a proper replacement for insulation PU foams.

3.2 Chemical Blowing Agents

It is well known that some chemicals are capable of liberating gaseous components via reactions and/or thermally induced decomposition. When these occurrences take place within the polymeric melt, the decomposing chemical automatically acts as a blowing agent. Some chemicals fit certain polymer processing nicely. These kinds of chemicals are referred to as chemical blowing agents (CBAs). As mentioned in the earlier section, CBA refers more to process than product. In comparison to the requirements set for the suitability of PBAs for foaming applications, the requirements for the processing suitability of CBAs appear to be more stringent. This is so because chemical reactions and/or heat are involved, so that the dispersion of the blowing agent throughout the melt and the heat sensitivity of the polymer impose serious concerns that aggravate the processing of polymeric foams using CBAs. In other words, heat sensitive polymer and the required shear to attain dispersion are legitimate material and processing issues. Moreover, common CBAs possess a decomposition temperature 100°C above the melting point of the semi-crystalline polymers. Removing the extra heat usually becomes a serious processing bottleneck.

The decomposition of a CBA not only depends on the processing thermal profile, but also on its residence time under the decomposition temperature. If it requires too high a temperature to trigger its decomposition, or takes too much time to complete the decomposition reaction, it will be extremely difficult to incorporate to the plasticator. A typical thermal decomposition curve is presented in Figure 3.1. The additional issues associated with selecting CBAs are listed in Table 3.2.

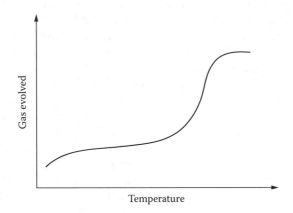

FIGURE 3.1
A typical chart for chemical blowing agent decomposition.

TABLE 3.2

Additional Considerations for Chemical Blowing Agent (CBA)

Processing compatibility with base polymers (decomposition temperature)
Kind and amount of decomposed gas(es)
Appropriate decomposition with other additives (i.e., peroxide)
Nucleating effects out of decomposed particles
Color from leftover or by-products

TABLE 3.3

Methods to Enhance Polymer Strength

Mechanism	Results	Application
Heat Dissipation/Solidification	Solidified Cell Wall	Foam Extrusion
		Injection Foam Molding
Chain X-linking, or Extension	Extensional Viscosity	X-linked PE Foam
		Low-Density PU Foam
Polymerization	Polymer Strength	Thermoset PU Foam

Most chemical reactions can either absorb heat or liberate heat, depending on the entropy summation of reactants vs. products. In contrast, when products become more active, it is generally required to add heat to proceed with the reaction. This type of reaction is referred to as being endothermic in nature. When the products possess less enthalpy than that of the reactants, it is exothermic. However, actual reactions often involve a primary reaction and a secondary reaction, such as further decomposition, or reaction between the reactant and the primary product. Net enthalpy balance is required to determine its thermal nature.

In fact, quite a few common CBAs are exothermic in nature. Exothermic reactions can promote gas expansion, but at the expense of weakening the polymeric melt strength due to the additional heating. At low expansion, polymer strength is usually not a concern, but strength becomes more critical as expansion ratio increases. Because the enthalpy plays a mixed role in the homogenizing, expansion, and stabilization stages of the foaming process, caution has to be exercised when selecting the CBA and foam fabrication method.

Most CBAs involve simple gases, which are very volatile. The inherent low solubility of the gas in the polymer combined with a high temperature profile may result in over-expansion and cell opening followed by collapsing. In contrast, PBA has less volatility and higher solubility to allow more dissolution in the melt. As a result, there are generally substantial plasticizing benefits, which renders more heat removal from the gas/melt possible. However, the volatile and less soluble CBAs virtually make a narrower foaming window. The density of the obtained foams would thus be relatively high. However, the volatile gas facilitates a sharp nucleation. If kept under control, it would be easier to obtain a fine-celled structure when using a CBA. The low expansion and fine-celled structure characterize CBA-blown foams, which become more suitable for batch or continuous mold-foaming processing (i.e., structural foams).

3.3 Reactive Foaming

As described in Section 2.4, gas molecules can be dispersed in the polymeric matrix by dissolution in the molten state for subsequent foaming and form- ing, or by reaction within the polymer prior to, during or post polymeriza- tion. In this context, Section 3.1 has been devoted to the dissolution of PBAs into the polymeric melt under high pressure and high temperature, whereas in Section 3.2, the CBA decomposition reaction which occurs within the melt to form the gaseous phase that creates the foam through pressure reduction induced super-saturation was analyzed. In both cases, conventional polymer processes, such as extrusion in free expansion and injection molding in constrained foaming, have been widely used in generating cellular structure for a variety of applications [87–88].

Clearly, gas formation within the polymer is a necessary condition for foaming, whereas equally or more important is the sufficient polymer capa- bility to hold the gaseous bubbles until a stable structure is developed. The role of the material strength increases as expansion ratio increases. When expansion exceeds 40 times, polymer strength almost dominates in the proc- essing. Although foaming could be very violent and far from an isothermal process, the polymer's capability should be high enough to control the volatile bubbling to ensure a stable foaming process. As tabulated in Table 3.3, there are at least three mechanisms to enhance polymeric strength for high-degree foaming. These include solidification, chain x-linking or exten- sion, and polymerization.

On the one hand, when polymeric strength is too high, the bubble expan- sion could experience too much existence to cause limited growth. A fine- celled structure can then be established, but at the expense of restrained expansion. On the other hand, when polymer lacks strength, excessive expansion quickly occurs. As a result, over-expansion, or bubble ruptures

TABLE 3.4

Material Considerations for Thermoset Foaming

Foaming	Foam Formation	Key Functions
Blowing Agent	Solubility in Reactants Reactivity in Reaction	Henry's Law Constant Permeability
Reactants	Functional Groups Reaction Rates Heat Evolved	Into Polymer
Polymer	Reactive Formation Proper Melt Strength	Propagation and X-linking Cell Structure for Property Solidification Permeation for Application

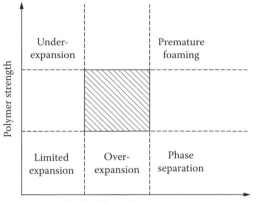

FIGURE 3.2
Foaming window; polymer strength vs gas/melt equilibrium constant.

are easily developed because of the weakened cell wall and/or broken structure. Sometimes, a nice expansion may occur, but, due to lacking of material strength, it will shrink sooner or later to a point where enough support is secured. For a polymer with a low modulus, both weakened and broken cell wall can contribute to foaming stability concerns because of which poor expanding behavior could be often experienced. The key concerns in foaming and foam formation are summarized in Figure 3.2.

Because CBA decomposition has been addressed in an earlier section, it would be appropriate to focus here on reactive thermoset foaming. In thermoset foaming, a tremendous material strength can be built up via polymerization and chain extension or cross-linking (x-linking) to accommodate the intensive (cell nucleation) and/or extensive (volume) expansion process. In fact, accurate timing of the competing reactions is the critical issue. In polyurethane foaming, the reaction necessary to generate the blowing gas and the polymerization/x-linking reactions intended to improve the polymeric strength are inter-related. The amount of gas formation, its diffusion, and the exothermic reaction heat dictate the expansion, and with a given number of bubbles, the bubble size can be readily calculated. The material strength required to hold the expansion can be estimated by considering the force balance expressed via Equation 3.1:

$$\eta = R_b/(dR_b/dt)\,(P_b - P - 2\sigma/R_b) \tag{3.1}$$

where R, P, and subscripted b represent radius, pressure, and bubble, respectively. As polymerization reaction continues, the viscosity increases, which favors holding gas expansion within the polymer. From the viscosity perspective, the gradual gas evolution out of the reaction becomes advantageous for foam. Sometimes, early gas may become the nucleus for the later gas to diffuse in. Polycondensation possesses the features for foaming and foam.

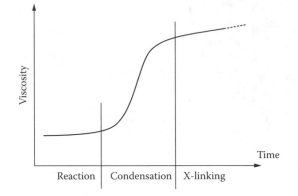

FIGURE 3.3
A typical viscosity change curve for thermoset polyurethane foaming.

In PU foaming, carbon dioxide, which has a low solubility in the polyol and isocyanate is produced first. Swarms of seeds are thus generated in addition to the entrained air bubbles from mixing the low viscous reactants. The low viscous nature of the reactants virtually favors nuclei growing above the critical radius. As reaction continues, more gases evolve and polymerization proceeds. The exothermic heat evidently fuels the expansion, which is quite unique for polyurethane, as discussed earlier. When x-linking via poly-condensation is in place, the PU foam can possess enough strength not only to hold the gas bubbles, but also to allow open-cell structure without collapsing [89,90]. It is quite different from thermoplastic foam processes, in which closed cell is very desirable for mechanical properties and dimensional stability. Fast blowing-agent escape becomes an additional benefit for open cell, provided that dimensional stability is under control. The reactive foaming and strength build-up through poly-condensation are depicted in Figure 3.3.

During foam expansion, extension, more precisely bi-axial extension, prevails over shear, as depicted in Figure 3.4. In thermoset PU foam formation, the cell wall becomes thinner as expansion continues so that the expansion rate and final expansion dictate the polymer strength requirement. In general, the extensional viscosity affects expansion rate, and the final expansion ratio. A low extensional viscosity favors expansion, and a high extensional viscosity is considered as beneficial for the cell integrity and its distribution. The important foaming parameters for thermoset reactive foaming are tabulated in Table 3.4.

Also noted is that the fresh PU foam temperature could exceed 130°C initially [91]. Fortunately, polymer is a poor thermal conductor to trap the heat in the core. As it begins to expand, the core becomes much hotter than the surface. HDPE film with melting around 130°C is sufficient to petition the foam from other contact without melting through. When foam eventually

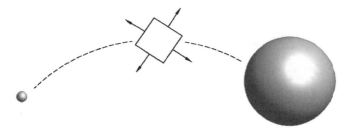

FIGURE 3.4
Surface extension during bubble growth.

cools back to room temperature, slight shrinkage due to gas volume reduction is anticipated.

3.4 Emulsion: Chemical Solution

Blowing agent is indispensable in the process of foam manufacturing, which leads people to ponder what is going to happen when polymer is overloaded with blowing agent. Polymer is no longer the dominant phase, and for certain it is a low viscosity solution. When people investigated the phase separation, it was found that according to phase diagram a continuous polymeric phase is possible at an overwhelming level of solvent (or blowing agent). A simple phase diagram with temperature as a ordinate is illustrated in Figure 3.5 [92]. When temperature is quickly dropped to solidify the polymeric fabric, a skeleton structure filled with solvent is obtained. After driving out the solvent through vacuum induced vaporization, a cellular structure ranges from open faced skeleton to closed cell polymeric matrix, similar to thermoset and thermoplastic foam. Depending upon the polymer property, it could be useful in specific applications. Biodegradable polylactic acid foam is a good example.

Of the few critical parameters in this methodology, one is the right polymer at the right weight percentage in the solvent. The second is the quenching, lowering the temperature to greatly under T_g or T_m to facilitate the solidification of polymer. Driving out the solvent is another critical component in this method. Quite a few polymers can be easily proved out in batch mode tests in the lab. However, a continuous process is always desirable from mass production perspective. Fortunately, linear polypropylene was found promising in dissolving in halogenated hydrocarbons to form a homogeneous solution at an elevated pressure and temperature. After cooling and foaming, a closed cell structure was achieved. This method was practiced in a continuous extrusion process to make an over 80 times expansion PP foam [93].

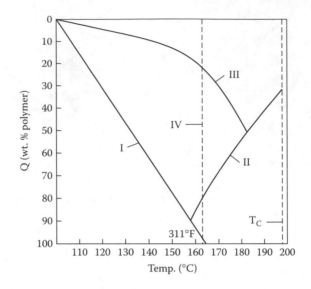

FIGURE 3.5
Phase diagram; weight of polypropylene in CFC-11 vs. Temp., I: solid/liquid phase line, II: sintering line (oriented), III: upper/lower operable line (fibrillation), IV: self-nulceation line, T_C: crtical temp. of CFC-11. [28].

    This solution method was further stretched. Polymer pellets were emulsified in a solvent. Instead of fully dissolved, the solvent can diffuse into and reside in the pellets. When saturation is reached, the pellets can be moved to a heated mold for expansion. This is the basis of molded bead foam technology. The concept seems quite straightforward: diffusion and expansion. But the process is batch in nature. Diffusion time, λ_D, is related to square of diffusion path, l.

$$\lambda_D \propto l^2 \tag{3.2}$$

The smaller the pellet, the faster the saturation. Also true is the heat conduction in inducing expansion. Conduction time is proportional to heat transfer path.

$$\lambda_T \propto l^2 \tag{3.3}$$

No wonder tiny bead in mm is preferred. That is where the beaded foam comes from. Adequate attention has been and is spent to fine tune the process. At present, it is quite popular in the packaging and food markets in using low density molded polystyrene foam and polyolefin foam parts. The closed and fine cell nature display excellent cushion and insulation properties.

Extrusion

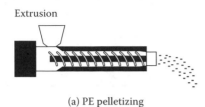

(a) PE pelletizing

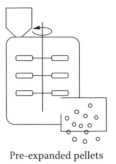

Pre-expanded pellets

(d) Drying

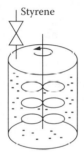

(b) PS polymerization/PE X-linking

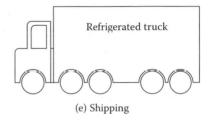

(e) Shipping

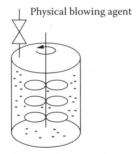

(c) Blowing agent impregnation

Mold

Product

(f) Forming and product

FIGURE 3.6
PE/PS Blend Foam Bead Schematics.

Most monomer is known as simple liquid under pressure. A blend can be made by emulsifying one polymer pellet in the other monomer. After polymerization is completed, a polymer blend is achieved. When blowing agent is introduced as shown in Figure 3.6, which is a molded bead process, we can obtain a blend foam.

In summary, emulsion appears to be another interesting approach to making polymeric foam. It involves simple diffusion and dissolution mechanisms to allow blowing agent into the polymer and/or disperse the polymer for expansion.

3.5 Nucleation in Batch, Continuous, and Reaction Foaming

This section addresses bubble nucleation, gas bubbles emerging from dispersed phase to clustered phase or non-spherical to spherical shape for expansion, in three different foam processing approaches: batch, continuous, and reaction foaming. Either thermal or mechanical super-saturation or both mostly cause bubble nucleation. Typical examples are thermal effects and/or rapid pressure reduction. Sizable research and development efforts have been devoted to spinodal decomposition and nucleation for polymeric foaming in the past decade, in which stable, meta-stable, and unstable states are generally involved [94]. As pointed out before, gas molecules respond to the change of surrounding conditions more obviously than the polymer. Basically, foaming is a way to dissipate change, or a response toward change. Gas molecules can diffuse into clusters to form a spherical nuclei, or, at a high concentration, can coalescence into a bubble phase known as spinodal decomposition. Resin crystallization [95] and polymer blend [96] identify spinodal decomposition and nuclei growth, which can be applied to foam formation as illustrated in Figure 3.7.

In general, spinodal decomposition and nucleation are distinguished mechanisms in phase separation. At high gas concentration, gas phase can group into phase separation with minimum transition, whereas at low gas

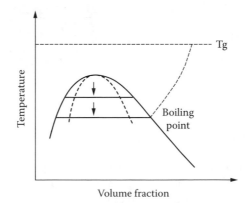

FIGURE 3.7
Temperature and volume fraction diagram to illustrate polystyrene foam phase separation.

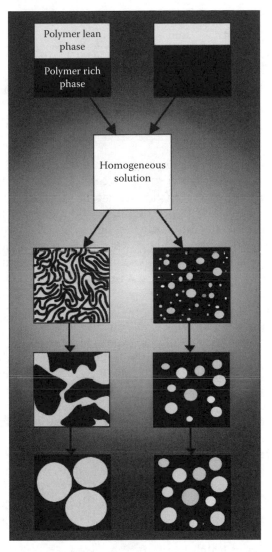

FIGURE 3.8
Spinodal and metastable nucleation.

concentration it takes time to allow adequate diffusion to form clusters known as meta-stable state, which, in turn, expands into spherical bubble. A general comparison between spinodal decomposition and nucleation is presented in Figure 3.8. In thermoplastic foaming, most systems belong to nucleation owing to low gas loading.

The batch foam processing assumes minimum polymeric melt motion so that nucleation is primarily controlled by thermodynamic laws. Molded bead [97,98] and batch microcellular foaming [59] are good examples. However,

TABLE 3.5

Summary of Nucleation in Different Foaming Processes

Nucleation type	Super-Saturation	Nucleation Mechanisms	Examples
Batch	Heat Induced	Homogeneous	Molded Bead Microcellular
Continuous	Pressure Drop	Heterogeneous Cavitation	Extrusion Injection Molding
Reaction	Over-Saturation	Turbulence/Entrainment	PU Foaming

in continuous foam processing, such as foam extrusion and injection molding, the mechanical energy (shaft rotation) converts to the motion of material (pumping), during which, due to viscous nature of polymer, inherent viscous dissipation generates heat within the polymeric melt, which unfortunately is done non-uniformly. It simply adds complexity to the already quite complex system. To reduce the complexity, it makes sense to begin with batch process. In batch reactive foaming, as soon as the gas generation exceeds the solubility limit of the surrounding mediums, foaming occurs inside or on top of the reactants. Apart from mixing induced bubble entrainment and phase inversion induced cellular skeleton, a positive superheat is necessary to generate spherical vapor phase within the polymer. In this context, we begin with a brief summary on nucleation types as presented in Table 3.5.

Clearly, in polymeric foaming, bubble nucleation and growth are simply responses to a sudden change of surrounding conditions. Both are dynamic in nature and occur regardless of the polymer being in motion or not. In other words, nucleation is a thermodynamically controlled rate process. Due to the dramatic difference in response to the condition change, the homogeneous melt turns into a heterogeneous system. As foaming goes on, the gas concentration gradient across the cell wall decreases. The activity of the system decreases, and the stability increases. But any disturbance such as motion induced momentum transfer and shear energy input can reverse the stability trend. It is therefore fitting to begin with a motion-free batch process.

3.5.1   Nucleation in Batch Foaming Processes

With respect to the static batch foaming process, the driving force is the superheat, or degree of super-saturation, defined as in Equation 3.2:

$$SH = P_{eq} - P \tag{3.4}$$

where P_{eq} and P denote the equilibrium pressure and the surrounding pressure, respectively.

$$P_{eq} = K_w W \tag{3.5}$$

where K_w and W are the Henry law constant and the weight fraction (i.e., the solubility) of gas in the polymeric melt, respectively. It is important to note here that after an adequate amount of time has elapsed, a homogeneous equilibrium will be established.

There are two common ways to generate positive superheat, i.e., by raising the system temperature or by reducing the system pressure, or by both. Commonly, heating and application of pressure reduction (or vacuum) are typical methods of bringing bubbles into existence [99,100]. The holding forces are the surface tension and viscous surrounding to constrain the expansion. Considering the surface tension and superheat as the dominant nucleation mechanism, the conventional homogeneous nucleation rate equation is thus developed:

$$J = A \exp (-B) \tag{3.6}$$

where, A and B represent material and system parameters, respectively. Virtually, B covers the driving and holding forces. A detailed expression is given by Equation 3.5 [101]:

$$J = N(2\sigma/(\pi m))^{1/2} \exp(-16\pi\sigma^3/(3kT(P_b-P)^2)) \tag{3.7}$$

where, N and m denote the number of molecules per unit volume of the metastable state, and the mass of a gas molecule, respectively. K is the Boltzman constant.

Figure 3.9 shows the trends in a log-linear chart. It is noted that the contribution in the system parameter, B, tends to have a straight line mode in the log-linear plot. In microcelluar foaming, qualitative agreement is observed in the parameter study [102,103]. However, quantitative agreement seems very difficult to establish. Accurate physical properties in the presence of blowing agent, other than the surface tension were investigated in earlier studies to achieve quantitative agreement [104–106]. Recent work based on the measured surface tension of the polymer saturated with the gas also indicated that a sizable gap still exists between the theory and experimental results. It is fair to say that the key parameters in most foam nucleation are attained, but their influence and at what stage are to be explored.

Because polymers, either in a molten state or near the glass transition temperature (T_g), have a long-chain molecular structure, which automatically causes substantial resistance for gas bubbles to expand and thereby represents a hindrance for the diffusion of gas molecules into the gas clusters, it is noted that a meta-stable state exists before the unstable growth state, which characterizes nucleation in most polymeric foaming processes. The transient meta-stable state can be shortened by an external energy input similarly to adding a boiling stone to induce dynamic water boiling when meta-stable boiling occurs.

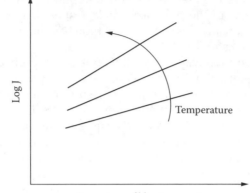

FIGURE 3.9
Nucleation at different gas dissolution.

3.5.2 Nucleation in Continuous Foaming Processes

In continuous foaming processes, bubble nucleation is accompanied with the motion of polymeric melt to render mutual effects between thermodynamics and kinetics in a subtler manner. In foam extrusion and injection molding, a rapid pressure-drop inherent of the process usually satisfies the super-saturation criteria for foaming. Also noted in processing is that processing pressure must be kept higher than the vapor pressure of gas/melt to ensure homogeneity of gas/melt in the processor. The processing pressure is determined by equipment design, rate, and formula. It basically reduces to atmosphere when exited to the outside atmosphere. However, the inherent vapor pressure is determined by the amount of gas dissolved in the melt, and temperature. It is independent of surrounding pressure. But motion induces viscous heat dissipation, which can affect processing pressure and vapor pressure in different ways.

It should be pointed out that as the gas/melt leaves from rotating shaft into shaping area (e.g., die), the flow is controlled by pressure gradient, known as pressure flow. The pressure difference between both ends is like "potential," some of which converts to material displacement, and others as viscous dissipation. The latter could affect vapor pressure.

Before we move further, let us define the nucleation point, which occurs when:

$$P_s = P_{eq} \tag{3.8}$$

where P_s represents the system pressure.

In this context, Figure 3.10 shows the pressure curves and nucleation point for the foam extrusion process. Figure 3.11 clearly demonstrates the

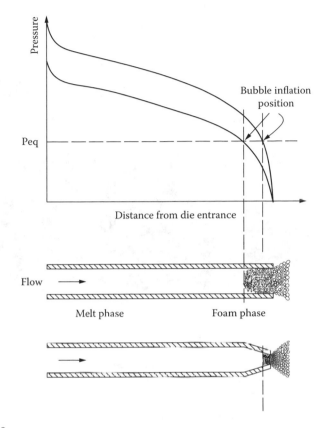

FIGURE 3.10
Foaming illustration for pressure flow in parallel plates; homogeneous to heterogeneous state.

nucleation points [107,108]. Namely, by considering the variations in motion, pressure, and temperature, it can be imagined that quantitative agreement of Equation 3.5 with the continuous foam process is extremely difficult to achieve. So far, only a semi-empirical approach, which is in qualitative agreement with the shear, was proposed [5,109,110].

In general, a pressure flow prevails at the onset of foaming (i.e., foam extrusion). In continuous foaming processes, two kinds of pressure drops should be distinguished: mechanical and chemical pressure drops.

$$\Delta P_{me} = P_s - P \tag{3.9}$$

$$\Delta P_{ce} = P_{eq} - P \tag{3.10}$$

where P_s represents the system pressure of flow, which is a function of position, or time, when the melt is inside the flow channel, and it is normally kept higher than P_{eq} to prevent two-phase formation. P_{eq} is determined by the amount of gas dissolved and the temperature.

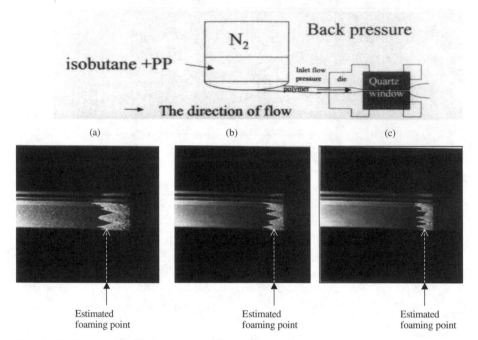

FIGURE 3.11

Whole fields of foaming behavior at various extrusion pressures: (a) 3, (b) 5, (c) 7 Mpa. (K. Taki et al., Visual Observation of Batch & Continuous Foaming Processes, Foams 2002, SPE, Houston, 2002. With permission.)

In certain continuous foaming processes, the superheat based on the chemical pressure drop underestimates the nucleation by orders of magnitude, whereas the mechanical driving force overestimates the nucleation rate as depicted in Figure 3.12. Even after fugacity (the actual vapor pressure over a vapor assumed to be an ideal gas obtained by correcting the determined vapor pressure and useful as a measure of the escaping tendency of a substance from a heterogeneous system) is introduced, the disagreement continues to exists [109].

When a blowing agent is dissolved in a processor, it tends to settle with an equilibrium state after enough time. In real plasticating, the melt is subjected to a smaller space to develop system pressure much higher than the equilibrium vapor pressure to render an under-saturation state, $P_s > P_{eq}$. In this state, the gas clusters cannot exist in spherical shape owing to aggressive surface tension force, they could only be in molecular tube or "string" form, or reside in cavity, such as surface of cosmic dust and contaminants. When the processor's pressure decreases, the coiled molecular tube tends to expand to cause contact to form a spherical gas cluster. Namely, after the onset of super-saturation, some curled tubes with enough molecules can transform into a spherical shape the radius of which is expressed by Equation 3.9

$$R_{cr} = 2\sigma/\Delta P_{ce} \tag{3.11}$$

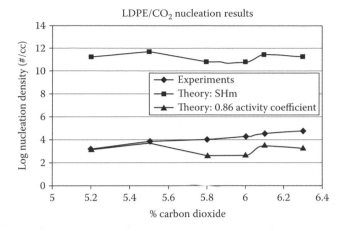

FIGURE 3.12
Nucleation for low density polyethylene with carbon dioxide [109].

where σ represents the surface tension. It is important to note here that P_s < P_{eq} is a necessary condition, sometimes not sufficient for bubble nucleation. It is certainly a meta-stable state. Any energy contribution can virtually lower the energy barrier for foam nucleation. A good example is to energize the system by shear or deformation can certainly facilitate the transform into the round shape because the shear energy has a straight line scattering in the log-linear nucleation plot [5,109]. This helps explain the shaking effects in foaming enhancement before or at opening a soda bottle.

It will be crucial to consider every energy parameter to establish the barrier-based nucleation kinetics. Conventional nucleation considers density fluctuation as the main driving force, and transport-controlled factors were later added to modify the equation as non-energy contribution. Therefore, molecular approach is quite necessary to establish a critical energy barrier, or critical super-heat, for nucleation.

3.5.3 Nucleation in Reaction Foaming Processes

When using simple liquids to form bubbles, the turbulence created by mixing the fluids entrains air as seeds for bubble nucleation. Similarly, in reaction foaming processes, the implemented way of mixing plays a critical role in foam nucleation in polyurethane foam. Namely, when the evolved gas exceeds the solubility limit, and when the diffusion into the existing seeds is not fast enough, more nucleation sites are developed. It thus appears that the mixing and reaction rate are the major nucleation parameters in the simple reactants foaming.

The urethane reaction is capable of generating tremendous amounts of gas and heat, which can easily bring the whole system into the spinodal decomposition regime. As reaction continues, newly evolved gas simply diffuses

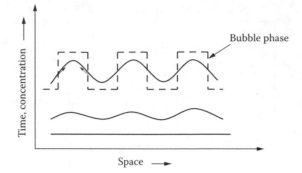

FIGURE 3.13
Spinodal decompsition in PU foam phase separation.

into the nucleated phase. Figure 3.13 illustrates the spinodal decomposition in thermoset PU foam formation. The number of cells is dictated by gas phase distribution, and the transformation of the long wave clusters. It was found that the seeded nucleus out of the usage of nucleator and entrainment has tremendous impact on cell density.

It is known that the total volume of gas bubbles is determined by the available gas phase in the polymer. Since nucleation followed by growth is of convolute nature, when bubbles start to expand, the interfacial area naturally increases and diffusion is enhanced. Then, the bubble growth rate is determined by diffusion, melt viscosity, and surface tension. However, since the total expansion is dictated by the amount of gas in the polymer, for a given intensity of dissolution (or reaction), nature of nucleation determines cell density and final cell size, more bubble and smaller size, or few bubbles and large size.

3.6 Growth

Gas phase that increases in size within the polymeric material is called "growth." A good example is spherical tiny bubble expands in polymeric melt. The bubble growth is essentially the main mechanism to dissipate the super-saturation state that is inherent to the polymeric foaming phenomena. It is known that after nucleation (tiny bubbles form and expand), the whole system is still in super-saturation. In fact, the degree of super-saturation appears to be the main driving force in most bubble growth. A number of references have been devoted to study the growth, ranging from a single bubble in an immense medium to a swarm of bubbles in a non-Newtonian fluid. In the early stage of growth, aggressive expansion on account of the instability occurs, which causes concentration decrease in the bubble to encourage diffusion. As the bubble continues to expand, the concentration gradient across the polymer and gas boundary decreases.

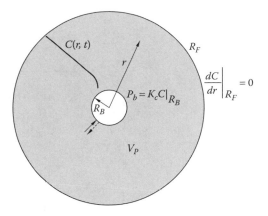

FIGURE 3.14
Cell model to describe bubble growth.

The so called "cell" model was proposed in the early '80s to investigate the growth of a swarm of bubbles in a finite medium [111,112], in which a shell of polymeric melt is assigned to a spherical bubble. It turned out to be especially useful for understanding the growth mechanism of thermoplastic foams. It assumes a spherical bubble enclosed by a shell of assigned polymeric melt as illustrated in Figure 3.14. Both mass balance and momentum equations are usually established to simulate bubble expansion. It was noted that the expansion involves mechanical expansion and chemical diffusion, which can be defined as follows:

$$P_{me} = P + 2\sigma/R_b \tag{3.12}$$

$$P_{eq} = K_w W \tag{3.13}$$

The above relationships indicate that the bubble pressure, P_b, should be greater than P_{me} for expansion in a viscous medium, whereas for diffusion it should be less than P_{ce}. In other words, diffusion and expansion are interrelated during bubble growth—cause and effect, and vice versa. Diffusion causes influx to bubble, and expansion tends to reduce gas concentration in the bubble for diffusion. The subtle criteria for growth is:

$$P_{ce} > P_b > P_{me} \tag{3.14}$$

These pressure differences become the driving force for expansion, which magnitude decreases as expansion continues. It can be imagined that at a small surface tension (or low viscosity), it is easy to expand, and diffusion proceeds at a relatively slow pace. It is called chemically controlled growth. However, at a large surface tension (or high viscosity), diffusion seems ahead to make it mechanically controlled. In theory, equilibrium will be established

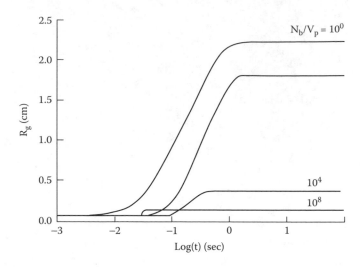

FIGURE 3.15
Simulation results for CFC-12 in polydimethylsiloxane with cell density as a parameter [113].

when the pressure differences disappear. A typical size expansion with different cell density is presented in Figure 3.15 [113].

However, in reality, bubble expansion takes place in a finite polymeric medium, and possibly in a different thermal environment. When foam exits into room temperature the natural cooling will solidify the surface to cause extra resistance for growth. In addition, gas molecules close to the boundary or surface can diffuse to the surface and escape from the surface, rather than into the cell as illustrated in Figure 3.16. Surface escape and the surface solidification effects can certainly decrease the foaming efficiency, which is defined as:

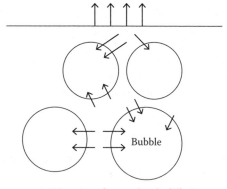

$\longrightarrow$ Direction of gas molecule diffusion

FIGURE 3.16
Gas molecules diffusion and surface escape in cell growth.

$$E_f = \text{Actual Expansion/Theoretical Expansion} \qquad (3.15)$$

Theoretical expansion assumes isothermal condition, no surface loss, and finite expansion. In fact, the foaming efficiency is a function of the geometry of the foamed product, or more precisely, its surface to volume ratio [114]. Owing to the surface tension force, the bubble size plays a critical role with respect to its pressure. Namely, the smaller the bubble size, the higher the internal bubble pressure is needed to sustain it. In other words, the theoretical volume is less for the smaller bubble than for the bigger bubble at a given load of gas. It is therefore necessary to define the foaming effectiveness as follows:

$$E_v = \text{Theoretical Expansion/Maximum Expansion} \qquad (3.16)$$

where "Maximum Expansion" is defined as expansion under the ideal gas law, or negligible surface tension. Figure 3.17 demonstrates the E_f and E_v in a simulation case. Consequently, it may be inferred that "efficiency" is a more geometry-inclined parameter, whereas "effectiveness" is more cell-structure-oriented. For example, bulky foaming is a highly efficient process, whereas fine-cell foaming has a low effectiveness. Consequently, thick microcellular foaming (cell size in 10 μm) is a highly efficient process that is characterized with a relatively low effectiveness. A summary chart is presented in Figure 3.18.

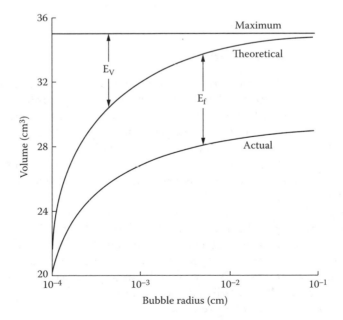

FIGURE 3.17
Maximum volume, theoretical volume, and actual volume vs. bubble radius.

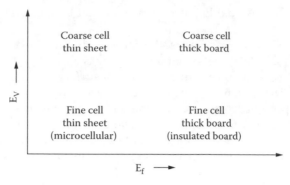

FIGURE 3.18
Cell structure in foaming effectiveness vs. foaming efficiency.

When the foam expands over four times its original volume, physical contacts between neighboring cells become inevitable according to the packing theory [115]. Once this occurs, bubbles of equal size, or spherical shape, in theory, cannot coexist. Instead, they can easily develop into multi-faced cell shapes, which further complicates their already very complex modeling. Nonetheless, the "cell" model allows us to investigate the material parameter analysis. The simulation results are especially important in the early growth analysis for experimentation [116].

3.7   Cell Coalescence and Rupture

Cell coalescence and rupture are naturally occurring during processing low-density foams (e.g., expansion over four times). When expansion increases, contact between neighboring bubble occurs, and thus cell wall sharing becomes inevitable. If the expansion continues, the shared wall will get thinner and thereby the cell wall will get weakened to a point where rupture may occur. In reactive thermoset foaming, drainage sometimes becomes another contribution to rupture. Basically, cell wall strength is the determining factor for rupture.

With respect to low-density thermoplastic foams, such as polystyrene and polyolefin foams, the cell contact is followed by cell distortion, as depicted in Figure 3.19, in which bubble contact causes distortion to make different curvatures. At high temperature and/or reactive thermoset foam, the drainage flow would occur to speed up cell wall thinning as a result of the pressure difference between the pressure in the corner (P_c) and the pressure in the cell wall (P_w) as shown in Figure 3.20.

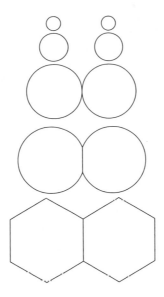

FIGURE 3.19
Bubble contact and subsequent growth.

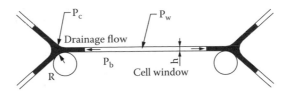

FIGURE 3.20
Drainage flow in bubble opening for polyurethane foam.

$$P_b - P_c = \frac{2\sigma}{R} \tag{3.17}$$

$$P_b - P_w = \frac{2\sigma}{\infty} = 0 \tag{3.18}$$

$$\therefore P_w - P_c = \frac{2\sigma}{R} \tag{3.19}$$

Eventually, when the critical thickness is reached, cell breakage occurs. When secondary nucleation is happening due to surface stretching during growth or reaching the cell growth threshold at a later time, bubbles with a smaller size have relatively higher internal bubble pressures. When these bubbles meet and establish physical contact, the expansion tends to move toward the "weak" side, which, in fact, is the bigger bubble [117], so that coalescence and rupture usually occur subsequently one after the other. In addition, in most thermoplastic foaming technologies, inorganic nucleating agents are being incorporated into the polymer melt. However, since these agents have minimum plasticity, when the expansion reaches a micron-scale level, it can be imagined that due to drastically different stretching capability between the inorganic and the polymer the interface tends to show holes. This causes the cell wall damage before it reaches the critical thickness, which, in turn, becomes open cells.

Cell wall ruptures create open-cell foams. Although this is desirable for certain absorption applications, such as fast gas-air exchange and absorption, it is detrimental for foaming efficiency [118] and the mechanical properties of the foam. For closed-cell thermoplastic foam applications, the latter is usually more critical. To reduce open cell fraction, melt strength enhancement is an effective material approach. It is known that branched polymers have been identified as having enhanced melt strength. Also, maintaining a reasonably low and uniform melt temperature during processing proved to be beneficial for the melt strength of the polymer. However, manufacturing foams with fewer open cells is a challenge for optimizing materials, equipment design, and processing conditions.

As shown in Figure 3.21, the cell wall rupture in thermoset foams occurs when the critical cell wall thickness is reached. Importantly, the cell wall begins to recede into strut-like skeleton which represents an open-faced structure, unlike the broken faced structure usually associated with thermoplastic foams. When the strut is strong enough, the open celled skeleton can be sustained and even withstand repeated compression/decompression for special applications.

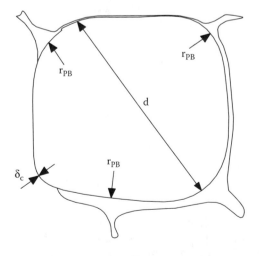

FIGURE 3.21
Cell Wall Opening. d, r_{PB} and δ_c denote cell diameter, curvature, and critical cell wall thickness, respectively.

3.8 Solidification and Curing

Both nucleation and growth are in essence dissipating energy to re-establish equilibrium. As pointed out before, the foaming process is unstable in nature. When volatile blowing agent or exothermic reaction is encountered, the foaming could be very dynamic. No matter how aggressive the foaming may be, as long as the surrounding polymer has enough strength to hold the foaming under control, the unstable foaming phenomenon may become a stable process. Therefore, sufficient material strength becomes a foaming parameter of critical importance because it ensures a stable process and useful cellular product. When foaming is completed, the polymer is either in a molten state or at elevated temperatures originating from exothermic reactions after which it cools into a solid stable state for applications. Cooling becomes the natural way to enhance material strength to keep the foamed product stable. Solidification and curing are referred to the cooling and air/ gas exchange stage. It is simply presented in Figure 3.22, showing the two main mechanisms: foam temperature and gas concentration as a function of time.

Foamed polymers are often used in applications exposed to the general surrounding atmospheric pressure and room temperature at which natural cooling takes place. Since polymers are poor heat conductors, and since the gaseous phase is even poorer, polymeric foams are inherently very slow heat dissipaters. However, the cooling time is strongly dependent on the size of the part, as depicted in Table 3.6 [119]. It shows that foam density and thickness are the primary parameters in characterizing heat transfer. Basically, it can take seconds to days until a thermal equilibrium between the foam and the surroundings can be reached. On the other hand, when the

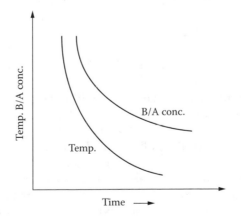

FIGURE 3.22
Decrease of foam temperature and blowing agent concentration during foam curing.

TABLE 3.6

Heat Transfer Calculations for Extruded PE Foam Sheet [119]

PE Foam Density, kg/m³	Thermal Diffusivity cm²/s x 10⁻⁴	Foam Sheet thickness, mm	Heat Transfer time, sec
100	1.9	1.6	135
30	2.8	1.6	91
30	2.8	0.8	23
30	2.8	0.4	6
30	2.8	0.2	1.4

foamed product is very thin, both the expansion and heat transfer times could be of the same order of magnitude. That suggests thermal effects during foam growth cannot be ignored. For instance, thin foam under 1 mm has cooling effects in foam growth.

Polymer conversion from a molten state to a solid state is a process of thermal nature. It is in fact the principal mechanism for improving the strength of the polymer and making it capable to hold the cellular structure. However, the viscosity build-up due to cooling evidently slows down the expansion and "freezes" the cellular structure.

In the meanwhile, during cell growth gas molecules close to the skin of the foam tend to diffuse and escape from the surface. This contributes to the loss of blowing gas, which, in turn, reduces the foaming efficiency [114,118]. Conversely, a lower foaming efficiency, or increased density, is often reported for high surface-to-volume products [117]. Considering the cooling, combined with the loss on account of the high surface to volume ratio, thin foam normally has a much lower foaming efficiency than that of the thick foam.

The polymer modulus curve is presented as a function of temperature in Figure 3.23 [120], in which the shape of the curve varies for the type of polymers. For amorphous polymer, or at above T_g, the modulus can be shifted along the logarithmic time scale. According to Williams-Landel-Ferry (WLF) equation, the well-known shift factor is:

$$\text{Log } a_T = 17.44 \, (T - T_g)/(51.6 + (T - T_g)) \tag{3.20}$$

As temperature continues to decrease, modulus tends to shift higher exponentially. Therefore, solidification is crucial to create enough material strength for holding the gas bubbles and imparting suitable mechanical properties for corresponding applications. PS foam is a good example.

In thermoset foams, the material strength is established by virtue of completion of polymerization and x-linking. Since inter-polymer bonds are developed, thermoset foams can allow excessive expansion to create open cell structure without causing concerns with regards to dimensional stability. Cooling back to room temperature is usually to increase the strength of the foam if intended for demanding applications. It should be noted here that

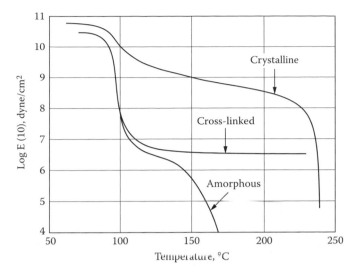

FIGURE 3.23
Elastic modulus variation as a function of temperature for polystyrene. (Adapted from A.V. Tobolsky [120].)

the influence of the surface tension decreases during the melt-to-solid phase transition, and eventually disappears when the solid phase is formed.

To avoid possible uneven stress build-ups during cooling of bulky foamed products, it would be desirable to maintain an even and symmetric heat dissipation. Controlled cooling appears desirable over natural cooling. Cooling in the mold is easier to achieve symmetric heat dissipation. But higher pressure associated with mold makes it fit for low (typically less than 1.5 folds) expansion products. That explains why structural foams, in general, have expansion much lower than the free expansion process.

As soon as cooling begins, the replacement of the blowing gas contained in the foam cells with the surrounding gas (i.e., air) begins. It is driven by the concentration gradient existing between the inside and the outside of the foamed product. Unlike the thermal nature of the solidification process, this is a permeation-controlled process. Typical gas concentration charts for both extruded and reactive foaming are illustrated in Figure 3.24. Blowing agent pressure decreases and air pressure increases as time passes by. For extruded foams having a very low density, the cell wall becomes too thin, so that it becomes responsive to cell pressure, which can create vacuum (pressure under 1 atm) as a result of fast permeation. In this context, a viscoelastic model was adopted to describe the permeation-induced foam dimensional stability phenomena, as illustrated in Figure 3.25 [121]. Characteristic loss of thickness and recovery over time are indications of the differences in permeability between the blowing gas and air. The ideal blowing gas should possess a near or same air permeation through the polymeric matrix during the curing stage.

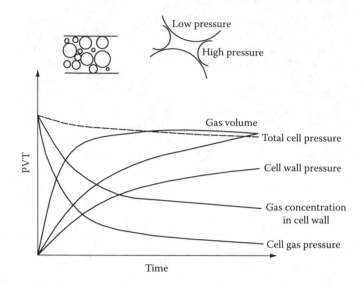

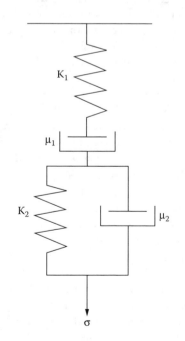

FIGURE 3.24
P-V-T variation during foam aging.

FIGURE 3.25
Maxwell and Voigt model to describe cell wall viscoelastic behavior.

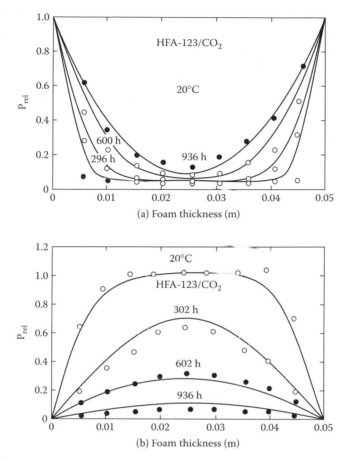

FIGURE 3.26
Concentration profile across the thickness of a polystyrene foam expanded with HCFC-123/ CO_2 [122]. (C.J. Hoogendoorn, Thermal Aging, in *Low Density Cellular Plastics*, Hilyard, N.C. and Cunningham, A., Eds., Chapman & Hall, London, 1994.)

Extruded bulky foam generally needs fabrication to desired shape and geometry. Generation of fresh surfaces becomes usually inevitable. When foam is not properly cured thermally and physically, residual stress and blowing agents can cause gross concerns. When nonsymmetrical fresh surfaces are generated as a result of the foam fabrication process of low density and low modulus foam (i.e., LDPE foams), the uncured foams tend to demonstrate unbalanced permeation, which, when large enough, can cause distortion and warping.

Although the curing process, i.e., the transition from a polymer/blowing gas-filled to polymer/air-filled cellular structure, begins as soon as the solidification starts, its completion can occur much later than that of the solidification itself, depending on the dimension and the permeation characteristics of the blowing gas. Thus, before the curing is completed, the blowing gas and the air co-exist in the polymeric foam cell. Figure 3.26 [122,123] illustrates the curing curves for polyurethane foam. However, it seems that there are reasons to believe that the blowing agent plays a transitional role, or acts like a "catalyst" during the curing process. Since often the thermo-conductivity, compressibility, and flammability properties of the blowing gas and the air are quite different, the residual content of blowing gas becomes a parameter that determines the thermal and mechanical properties of the foam.

4

General Foam Processing Technologies

Foaming within polymer is a unique and intriguing phenomena. It is dynamic in nature, and the pressure-temperature-volume varies throughout the process [123]. A stable and useful foamed product is formed at the end. It can be imagined that from concept to lab success, from lab to pilot line, and from pilot to commercial success, involves a wide variety of challenges. It is very encouraging that the foam industry not only survived the challenges, but picked up strength after each trial to become stronger. In this chapter, we will address several established foam technologies to comprehend the engineering practice of the scientic principles.

4.1 Foam Extrusion

An extruder, consisting of a rotating flighted shaft in a stationary barrel, is a well-known plasticator for its inherently efficient and effective energy transfer and positive displacement. When a thermoplastic polymer has a fitting processing latitude with enough difference between the melting and decomposition temperatures of polymer, extrusion is a useful fabrication process to turn raw material into useful products. The screw is deliberately designed to have a high contact surface-to-volume ratio, and therefore extrusion is ranked as a plastic fabrication process capable of generating the most efficient heat transfer. Consequently, in extrusion, the molten state of the polymer can be achieved within a reasonably short time without causing any thermal degradation. Extrusion equipment designers found that the extrusion efficiency can be improved by varying the shaft and barrel diameter ratio. In addition, the internal pressure can be controlled by improving the design of flights and shaft dimension, which is especially important in introducing gaseous blowing agent into polymeric melt.

Modifying the performance of the base polymer by adding another polymer or additives with a view toward enhancing its application becomes a common practice in extrusion. Conversely, adding blowing agents into the molten polymer should be regarded as a natural extension of the

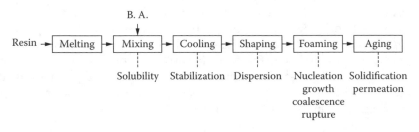

Gas
↓
Solid → Melt ---- Gas/melt solution → Bubble/melt → Bubble/solid

	Extrusion		Foaming	Aging
Low temp.	High temp.		High temp.	Low temp.
Low pres.	High pres.		Low pres.	Low pres.

FIGURE 4.1
State change in a typical foam extrusion process.

B. A.
↓
Resin → | Melting |→| Mixing |→| Cooling |→| Shaping |→| Foaming |→| Aging |

Solubility Stabilization Dispersion Nucleation   Solidification
                                       growth   permeation
                                       coalescence
                                       rupture

FIGURE 4.2
Foam extrusion units and their mechanisms.

conventional extrusion [124]. Figure 1.8 shows a common tandem line foam extrusion system.

A typical PBA-based foam extrusion consists of polymer melting, polymer/gas mixing, cooling, and shaping. As outlined in Figure 4.1, it is primarily a state change process during which the blowing gas is initially dissolved throughout the polymer matrix to subsequently form spherical bubbles while involving a variety of thermodynamic, kinetic, rheological, transport, and design issues. The necessary units for foam extrusion are melting, mixing, cooling, and shaping, which along with its relevant mechanisms are outlined in Figure 4.2.

The screw revolution is virtually a constant source of mechanical energy, which can transfer to the polymer as a pumping as well as heating source via friction and viscous dissipation. As the polymer moves toward the downstream, the screw can be designed in such a way that a cross channel velocity exists to promote mixing. That is essentially necessary for both the melting and mixing purposes, whereas it becomes a critical issue with respect to heat removal which becomes necessary to stabilize the foaming system. Creating an appropriate design to cover each mechanism involves more art than science [125,126]. Nonetheless, through proper design of the screw and seal, extruder can be regarded as reactor and heat exchanger, respectively, which greatly enhances its capability for mixing and cooling control.

The melting stage basically is to convert solid pellets into molten state. The main heating is out of barrel heat conduction and solid friction to change from full solid bed to solid bed/melt pool to full melt pool. A barrier screw

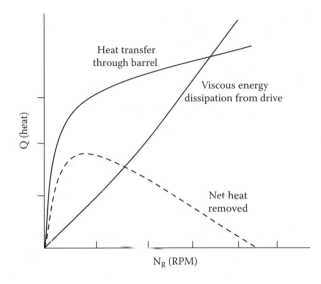

FIGURE 4.3
Cooling extruder characteristics; input work and heat removal.

is used to separate the solid bed and melt pool to enhance the melting efficiency. The next one is mixing. When gas is introduced to the polymeric melt, a distributive mixing device is necessary to break the gas stream into small pockets for fast dissolution. After the mixing, a homogeneous solution is expected, which moves to cooling stage. It is by far the most challenging unit in design of efficient commercial extrusion process. On the one hand, screw rotation continues to bring energy into the gas/melt. Heat removal from the barrel evidently requires effective surface renewal to reduce the bulk heat. There is a trade-off as shown in Figure 4.3. It requires expertise to achieve the optimal design for a high rate commercial process.

It is known that a hot spot exists in the channel, as illustrated in Figure 4.4, which is located in the upper center. That hot spot can easily render a

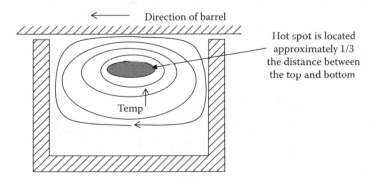

FIGURE 4.4
Temperature distribution in a typical melt channel.

foaming outside the acceptable window to fail the whole product. General modification is to introduce broken flights to disrupt the flow to break the hot region. It helps, but the downside is the decrease of pumping, and lack of vertical mixing [127]. The other practice is to reduce the polymer coating on the interior barrel, which results from barrel cooling. A small gap between flight tip and barrel is a good practice, which may reduce the lifetime of the screw. Other recent revisions include elongational flow, holed flight, and layer reunion [128,129]. It was reported to have over 2,000 kg/hr in PS board foam in a tandem system with the secondary extruder at 400mm diameter.

After cooling is done, the stabilized gas/melt is pumped into a shaping unit for foaming. Uniform velocity, and uniform thermal distribution are necessary for uniform foaming. In this stage, screw rotation is no longer existed. Static mixer and melt pump are general practice for uniform temperature and pressure for foaming.

As long as the unit operation is met, extruder with any screw: single screw, twin-screw (co- and counter-), and multi-screw, can be adopted and implemented for general foaming operations regardless of the existing number of screws and extrusion units. In general, the screw dimension and length to diameter ratio (L/D) determine the capability of all functions. Table 4.1 includes general information for various extrusion systems [34,35].

Since the 1950s, the continuous extrusion process attracted sizable development efforts for foaming polystyrene and polyolefins. Nowadays, it is a very popular technology for the manufacture of a variety of foamed products most of which became almost indispensable in our daily lives, ranging from converting polymers into thin foam sheets for thermoforming into a cup or tray to the manufacture of thick insulation boards. Thin sheet manufacture benefits from the high pressure associated with the smaller die opening for achieving lower-density foams, but high shear causes shear-thinning concerns. In contrast, for the manufacture of thick boards the lower processing pressure holds less dissolved gas to make relatively higher density foams, whereas the low shear is beneficial for improving the properties of the foam [130].

As indicated earlier, extruded products are limited by geometry. Roll stock production is a common practice, as shown in Figure 4.5. Slab, sheet, rod, and profile are also possible from the extrusion process. In general, the foamed products go through a festooned process, or conveyer system to be

TABLE 4.1

General Information for Foam Extruder

Type	L/D	Products	Remarks
Single Screw	Over 30	Rod, Sheet, Shell	Easy to control
Twin Screw	Over 25	Rod, Sheet, Board	Able to make board
Tandem Lines	Over 15 for primary Over 20 for secondary	Sheet, Board	Independent cooling control

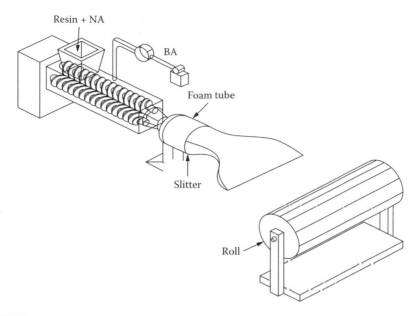

FIGURE 4.5
Schematics of twin screw foam sheet process.

collected for curing, during which they cool down to acceptable temperature for warehousing or for further fabrication. In certain processes, vacuum can be applied to enhance foaming after it exits the shaping unit. U.C. Industry suggested a vacuum box sealed by a water pool, which can also solidify the cellular structure [131]. The main benefits are more expansion, and fine surface structure.

However, post extrusion fabrication is necessary to convert the foam into useful products such as pouch, mailer, and bag. Lamination is another common practice for foam. PS sheet is bonded to barrier film for meat, poultry, and vegetable freshness. Specialty films can certainly enlarge the property spectrum of the foam products. Aluminum foil, permanent antistatic film, high friction film, corrosive inhibiting film, etc. can laminate to foam for a variety of applications.

Thermoforming is another way to convert the flat sheet into a final geometry. Trays, cups, and shoe soles are good examples for PS foam, PP, and PVC foam. It expanded from food industry to recreation and the automotive industries in the last decade. The post extrusion fabrication continues to become more extensive to meet diverse market demands.

The constant drive in foam extrusion is not only to make a product, but to make a better product with improved efficiency. It is basically a self-upgrading process. Cooling screw design to enhance the heat exchanger efficiency is one aspect. It certainly needs improvement from other areas as well. For instance, a more forgiving gas/melt system, an integral extrusion/fabrication system for the final product, and a more efficient change over system from product to product. When processing and processor have sufficiently improved, the use of volatile inorganic blowing agent in extrusion could be possible by then.

Reactive extrusion was recently considered as a viable way of improving polymer foamability by modifying the base resin in-situ to enhance the extensional strength which is essential for holding the gas. In this context, branching, x-linking, and grafting side chains are regarded as effective ways to enhance the foamability of PET and PP [132,133]. In summary, formula, processing, design, and fabrication represent the fundamental technical requirements for improving foam extrusion.

4.2   Thermoset Reactive Foaming

Thermoset reactive foaming basically includes foaming via reactants to liberate gaseous components, and polymerization to encapsulate the gas phase. It is different from the thermoplastic reactive foaming, which has heat activated decomposition of CBA within a thermoplastic matrix. It could be accomplished in extrusion, injection molding, and oven heating processes. This section is dedicated to the foaming of thermoset polymers, and thermoplastic reactive foaming will be addressed in Section 4.4 entitled x-linked PE foams.

Polyurethane, polyisocyanurate, phenolics, and epoxy are the popular thermosets. Polyurethane is the most well-known in thermoset foaming. It is basically made by mixing polyols, isocyanate, amine catalyst, and additives to allow urification and PU-merization reactions to happen to evolve gas and heat for foaming.

The processing of polyurethane foams relies on using reactants to achieve proper mixing and foaming to a desired expansion and/or geometry. Both the continuous and semi-continuous polyurethane foaming processes, as well as the batch molding method, have been established for decades in the polymer foaming industry. Polyurethane foams are commonly made through a reaction between isocyanate and polyol streams in the presence of catalyst and additives for immediate foaming reaction. TDI (toluene diisocyanate) and MDI (diphenylmethane diisocyanates) are typical examples for isocyanate. However, the intended application of the final product strongly dictates the selection of the manufacturing process. Therefore, both free expansion and restricted expansion methods can be implemented to tailor the intrinsically rich volume expansion of polyurethane foams and thereby satisfy specific application requirements. In this context, a brief summary of each process is presented in Table 4.2. From the product perspective, it can be categorized as: low density flexible foam, semi-rigid, and rigid PU foam. The products and processes are summarized in Table 4.3.

A horizontal conveyer belt with mixing chamber in the front, into which nozzles carrying chemical streams are fed, as illustrated in Figure 4.6, is a

TABLE 4.2

Advantages and Disadvantages of Different Reactive Foaming Processes

	Advantages	Disadvantages
Continuous Process	Maximum Expansion High Speed No Mold	Uneven Expansion Need Fabrication Longer Cooling
Semi-Continuous Process	Good Yield	Less Efficient Processing
Batch Process	Foaming in Mold Restrained Foaming	Mold, Design/Maintainance Not Maximum Expansion

TABLE 4.3

General PU Foam Products and Processes

Products	Processes
Low density Flexible PU Foam Batch Free Expansion Process Batch Mold Process	Continuous Slab Processes
Medium Density Flexible PU Foam	Batch Mold Process
Semi-Rigid PU Foam	Integral Skin Process
Low Density Rigid PU Foam Mold Process (Vacuum Insulation Panel)	Continuous Conveyor/Oven Process
Rigid PU Foam Free Expansion Spray Foam Molded Structural Foam	High Density Mold Process

typical process for flexible PU foam. Uniform expansion can be controlled by catalyst and temperature. Sometimes cooling is necessary at the mixing, and heating for expansion. It can be modified to allow foaming flow going upward and set before the exit of a vertical channel. Also found was that gentle expansion as illustrated in Figure 4.7 can be used in the packaging application, in which PU foam fills the void to compact the packaging for transportation. It is a common practice in medium weight product protection.

In general, PU foaming is not very aggressive and is suitable for mold filling. Naturally, it found in-roads to mold foaming processes. More details will be provided in Section 4.3. A common practice is known as reactive injection molding (RIM) technology. In mold filling, mold temperature can be controlled in such a way that surface foaming collapses into skin at contacting hot mold with internal foaming acting for extra crushing to make an integral skin semi-rigid PU foam. As for rigid PU foam, a rigid frame is pre-set for filling, foaming, and setting. Appliance door as shown in Figure 4.8 is a typical process. Since vacuum offers excellent insulation, vacuum insulated panel (VIP) is an interesting foaming and vacuuming methodology for PU foam.

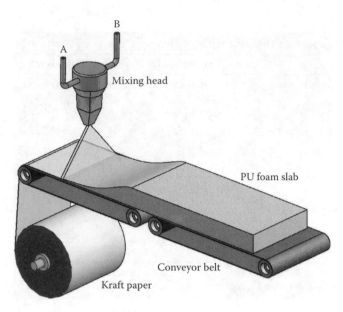

FIGURE 4.6
A simple schematics of a continuous horizontal PU foaming process.

FIGURE 4.7
PU foam expansion experiment.

4.3   Foam Injection Molding

Injection mold has become a very popular plastication technology, owing mainly to the diverse mold design and wide range of ram tonnage. Gas involved in plastic mold processes has a long history [134]. So far, the following five processes are utilized by industry for various reasons:

1. Reactive Injection Mold (RIM) foaming
2. Low Pressure Structural Mold foaming

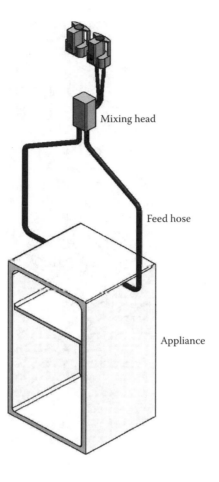

FIGURE 4.8
Rigid PU foam for appliance door.

3. Co-injection Structural Mold foaming
4. Gas-Assisted Injection Molding
5. Microcellar Injection Molding

Reactive injection mold (RIM) foaming has simple chemical fluids, such as polyol, and isocyanate with catalyst, pumped into a mixing chamber for preliminary reaction. It is then injected into the mold for foaming, filling, cooling, and setting. After being ejected from the mold, the foamed product is basically ready for use. Figure 4.9 illustrates this popular process which is common in PU foam industry.

In injection molding, the polymer melting and mixing operations resemble very much those in extrusion. When the gas phase is implemented in the molten polymer, either through CBA decomposition or PBA dissolution, the homogenized melt is pushed by a ram into a thermo controlled mold for subsequent foaming, forming, and, later, internal cooling and ejection. Figure

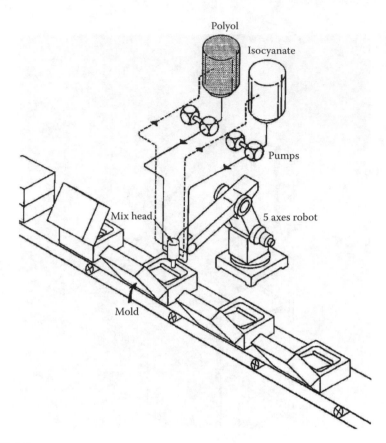

FIGURE 4.9
Schematics of PU Reactive Injection Mold (RIM) foaming.

4.10 shows a typical schematic of injection mold foaming for preparation of foamable composition. Unlike free foam expansion which usually occurs in extrusion foaming, the expansion in foam injection molding is constrained. The mold becomes the expansion limit, as opposed to the material strength in extrusion. In fact, these are the principal differences between these two foaming technologies.

A simple example of mold filling and foaming is depicted in Figure 4.11. At the entrance, the pressure drop causes the super-saturation necessary for foam nucleation. As the polymeric front progresses to fill the mold, cell growth occurs. However, the volume expansion of the foam is constrained by the size and shape of the mold. Since the temperature of the mold is generally kept lower than that of the melt, this can easily cause instant polymer cooling at the contact surface thereby forming a skin layer, whereas the remaining material throughout the mold's core still remains hot for foaming. A dual structure can thus be prepared. A common practice is the structural mold foaming with CBA. A low pressure mold is held at a low temperature to solidify the skin, and foaming occurs in the core to fill the

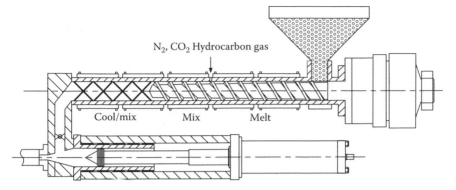

FIGURE 4.10
A typical injection mold foaming process.

mold. Sometimes, counter pressure is applied to prevent premature foaming. It should be noted that during mold filling, a dead spot can be generated in case an escape route for the remaining gas is not adequately available. As a consequence, rough surface spots may exist. A common remedy is to implement vent holes for a controlled pressure decrease. Expanding mold is another practice in structural foam to accommodate the expansion.

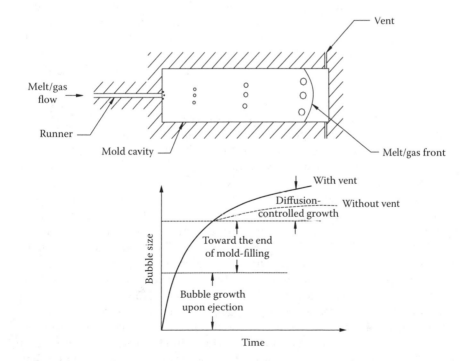

FIGURE 4.11
Mold filling and foaming for injection mold process.

TABLE 4.4

Comparison between Foam Extrusion and Injection Molding

Items	Extrusion	Injection Molding
Nature	Continuous	Batch or semi-continuous
Blowing Agent	PBA and CBA	CBA preferred
Foaming	Free Expansion	Constrained Expansion
Product Shape	Common	Common and Complex
Foam Density, g/cm^3	0.05–0.5	Over 0.2
X-linking	None to Slight	Slight to medium
Cell Distribution	Normal	Dual

The cell pressure can be controlled by controlling the amount of blowing agent and/or the size of the mold. It can be expressed by calculating the cell density and average cell size. In addition, the mold temperature can be controlled for foaming and curing, which affects the skin structure and the cell size of the foamed core. A simple comparison with foam extrusion is presented in Table 4.4.

Co-injection mold foaming draws more attention lately, because it involves a hot mold and the cycle time is less. Basically, skin and core are injected to the mold alternatively. When the skin is injected first, foaming can certainly help fill the mold to generate a compact skin-foamed core structure, or foamed skin-compact core the other way around. It shows a combination of different degree of foaming of the same material, or different polymers for desired property. Caution has to be exercised when compatibility of two different polymers becomes an issue. A simple schematic is presented in Figure 4.12.

Another interesting practice is the gas-assisted injection molding, in which an additional nozzle is used to introduce nitrogen gas at prescribed pressures at different processing stages. It can easily generate a gas space as large as 50% of the volume of the formed product. Table 4.5 demonstrates this comparison between gas-assisted and foam injection molding. The main advantages are no sink marks, and no or low warping [135].

In recent years, microcellular injection molding gained a wide usage in the injection mold industry. An inorganic PBA (e.g., CO_2 or N_2) is introduced to the melt in the mixing section to form a homogenized melt/gas solution, which is rammed into a low pressure mold. The final cellular structure possesses cell size in the magnitude of 10 microns. Reduction of cycle time and tonnage are reported as main benefits. It is also good for thin and small parts. A typical process is illustrated in Figure 4.13 [136].

In summary, the injection molding process seems to be better suited for foaming control than extrusion because it offers a better control of the pressure reduction and mold temperature. The skin structure and cell size can be better controlled. In general, it is ideal for performance polymers to have a finer cell and higher density. Recent reports on microcellular foaming in injection molding based on PBA injection revealed that the presence of foam

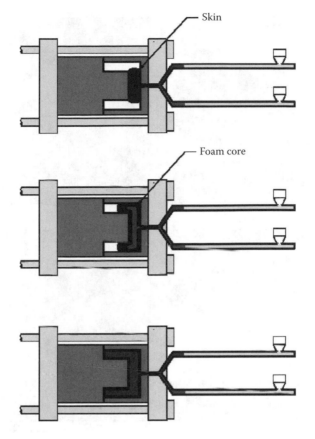

FIGURE 4.12
Co-injection molding schematics.

TABLE 4.5

Brief Comparison between Gas-Assisted and Foam
Injection Molding

Items	Gas-Assisted	Foam Injection Molding
Structure	Dual, skin/core	Homogeneous
Skin Structure	Smooth Skin	Fine-celled smooth skin
Core Structure	Void	Cellular structure
Fast Cycle Time	None	20–50% reduction
Material Saving	30–70%	20–60%

speeds up both mold filling and part cooling, which greatly reduces the required cycle time [136]. In certain cases, it outpaces the material saving proposed by microcellular foaming.

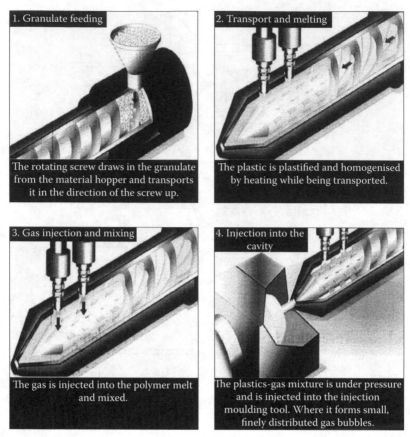

1. Granulate feeding

The rotating screw draws in the granulate from the material hopper and transports it in the direction of the screw up.

2. Transport and melting

The plastic is plastified and homogenised by heating while being transported.

3. Gas injection and mixing

The gas is injected into the polymer melt and mixed.

4. Injection into the cavity

The plastics-gas mixture is under pressure and is injected into the injection moulding tool. Where it forms small, finely distributed gas bubbles.

FIGURE 4.13
Microcellular injection molding (K. Okamato, *Microcellular Processing*, Hanser, Munich, Germany, 2003. With permission.)

4.4 X-Linked Polyolefin Foam

The x-linked PE foaming process represents another category of reactive foaming. It refers to build up bonding between polymer backbones to enhance material strength. In general, it is accomplished by chemical reaction or irradiation. Chemical reaction includes peroxide initiation to generate free particles, and silane reaction. Electron beam is a common practice in creating free particles for chain formation in solid state. Nowadays, peroxide and electron beam are well-established commercial processes, known as chemical x-linked, and irradiation x-linked.

In the chemical x-linked process, x-linking agent and CBA are compounded via extrusion or roll mills into PE sheets which are subsequently subjected to first heat to activate the x-linking agent, and to a second higher

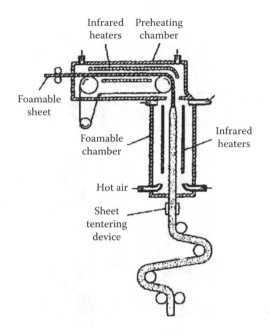

FIGURE 4.14
X-linked PE vertical foaming process [137].

heating zone to decompose the CBA for foaming. In general, peroxide and irradiation are good practices in generating free particles. A common outline for x-linking PE foam is presented in Figure 4.14 [137]. Typical x-linking processes are tabulated in Table 4.6. The main feature of X-PE foam is the softness and fine cell structure. Besides, its increased strength makes itself fit for thermoforming, which opened up more application opportunies.

In compounding, the main challenge is the even dispersion of CBA and peroxide in the polymer, and even thickness of the sheet. It has to be subjected to a low temperature profile to prevent CBA decomposition, which causes extrusion rate concerns. After the sheet is made, it will be exposed to an oven with temperature set at 100°C above the melting point of LDPE. The strength of the material to hold together at such a high temperature is very challenging. A specially designed loop/conveyor system is critical in a smooth operation.

Conventional linear polypropylene has a melt point above 160°C, which leaves minimum room for processing without causing CBA decomposition. However, polypropylene copolymer can be processed at a much lower temperature. In the presence of antioxidant, PP copolymer is successfully blended with CBA into a sheet form, which is foamed through the oven process for automotive applications.

Physical inorganic blowing agent has been incorporated into x-linked PE sheet, which is placed in a high pressure and higher temperature mold or chamber for impregnation. After impregnation is done, or saturation is

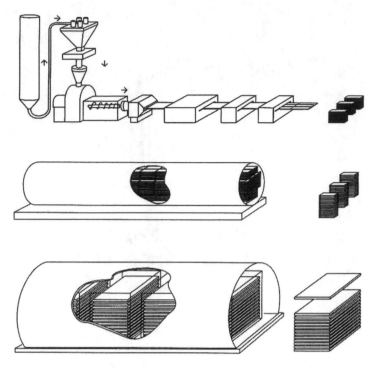

FIGURE 4.15
X-linked PE foam with physical blowing agent. (Courtesy of Zotefoam.)

accomplished, pressure can be released to promote foaming in the mold. X-linked foam can thus be made without using CBA. Figure 4.15 illustrates the X-PE foam slab with nitrogen blowing agent.

4.5   Molded Bead

Another well-established foaming technology is the molded bead method. Its development route is similar to that of foam extrusion. The initial success was reported with respect to foaming polystyrene in the 1960s, and since then it has been widely extended to polyolefin resins. It was speculated that its glass transition temperature and amorphous morphology are the key factors that allow successful polystyrene high-degree foaming up to over 40 times.

The molded bead foaming process is based on using resins soaked with a physical blowing agent at an elevated temperature and then exposing them to super-saturation at either a lower pressure, or a higher temperature, or both. A liquid medium is often involved to accelerate the soaking and/or foaming. Staged foaming is another useful practice in achieving optimal

TABLE 4.6

Summary of X-Linked PE Foam Methodology and Product Geometry

X-Linking Method	Method	Product Geometry
Peroxide	Horizontal oven	Sheet
Peroxide	Mold	Board, over 5 cm and 20X
Irradiation	Horizontal oven	Sheet, thin gauge
Irradiation	Vertical oven	Under 1 mm and over 20X
Silane	Oven	Sheet

expansion. Typical applications of polystyrene soaked with hydrocarbon include loose fill packaging, foam cups, trays, and end caps for protection [3].

A simple schematic of the processing principle of this technology is illustrated in Figure 4 16. The beads are subjected either to free form or constrained expansion. The former is known as "peanuts" for loose fill in packaging. The latter is the foaming in the mold. It consists of intra-bead expansion and inter-bead fusion. The main advantage is the fine-cell structure at a controlled expansion. In addition, this process inherently has minimum shear history, which is favorable for cell expansion and the preservation of the mechanical properties of the foam. Unlike the continuous extrusion process, the batch nature of molded bead foaming causes process efficiency concerns. A simple comparison with foam extrusion is presented in Table 4.7.

On the other hand, sizable efforts have been devoted to foaming polyolefin and its blends. It was found that x-linked polyethylene is capable of foaming into fine celled mold bead products [138], in which the enhanced melt strength was credited to the dimensional stability. However, when random polypropylene copolymer was successfully foamed [139], it was concluded that an adequate foaming window could be established for successful foaming of polyolefins. Thus, as long as the blowing agent loading and material strength fall into the right window, nicely foamed and stabilized cellular products can be expected, even for foamed alloy [140]. The low shear history imparted to the processed parts by this process may offer further benefits with respect to part properties, in comparison with those made by using the medium-to-high shear extrusion process.

4.6 Foamed Film

The extruded blown film process is a well-established method of making polyethylene film either with stalk (for linear PE) or straight blowing (for branched/linear PE). It is straightforward to blend a blowing agent into the

TABLE 4.7

Comparison between Commercial Foam Extrusion and Molded
Bead

Items	Extrusion	Bead
Formula		
Resin	Branched	Linear, Copolymer, X-linked
Blowing Agent	Hydrocarbon	Hydrocarbon, CO_2
Processing	Continuous	Batch
Temp.	High	Medium
Pressure	High	Low-Medium
Foaming	Free Expansion	Restrained Foaming
Equipment	None	Mold
Efficiency	Very High	Medium to High
Blowing Agent Loss	Some	None
Aging		
Permeation	Natural	Natural
Residual Gas	Low-medium	Medium to high
Product		
Density	Low to high	Very low to medium
Cell size	Sub mm-cm	Micron to mm
Property	Good	Better
Recycle	Yes	Yes
Reprocess and Reuse	Low	Medium to high

formula to generate a cellular structure in the blown products. With proper
adjustment of the processing parameters, a cellular structure can be introduced by adding a chemical blowing agent, either endothermic or exothermic in nature.

In general, a high processing temperature profile is necessary to decompose the CBA, which evolves gaseous components (i.e., nitrogen or carbon
dioxide). Their high volatility and low solubility prohibit the dissolution of
a large amount of gas(es) and, furthermore, the high diffusivity of gaseous
blowing agent will cause over-expansion. Therefore, the achievable expansion is generally much lower than that in PBA-based foaming. Endothermic
CBAs, such as citric acid and sodium bicarbonate, are preferred because they
absorb heat from the molten polymer during the reaction and are thus
favorable for successful foam stabilization [141]. In addition, a normal blown
film line has a relatively wide die opening to generate low shear and thereby
facilitate die flow for achieving the desired post stretch. The optimal die-
gap-to-film-gauge ratio is around 50:1. For foaming, the gap needs to be
closed down to suppress the evolved blowing agent in the melt in order to
prevent premature foaming. Also, it has been noted that the blown film
process allows for radial expansion, which is an advantage over cast film
with respect to optimal foaming. Nonetheless, the expansion is limited to
up to three fold. In conventional film processing, a high temperature profile

| Virgin beads with blowing agent (about 35 lb/ft³) | Pre-expanded beads (about 1 lb/ft³) | Molded expanded beads, (about 1 lb/ft³) or less (note that the expanded beads have fused into each other). |

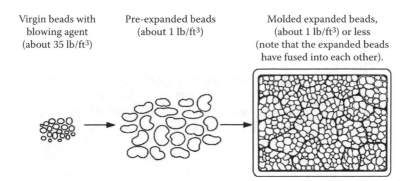

FIGURE 4.16
Molded bead schematics.

is normally kept for enabling flow distribution control. But, since foam processing is very sensitive to temperature and pressure, design revision is often necessary to achieve improved expansion and cell density control.

Although PBAs can be injected to create cellular structures in the blown film process, the limited expansion compared with other technologies renders the resulting material saving non-attractive. When other requirements, such as translucent instead of tranparant, cushioning, and thicker product, become prevailing. It will make sense to inject PBA into the blown film process.

However, it is a common practice in more layers co-extrusion film process for specific applications. In general, the core layer is a specialty polymer, such as polyvinylidene chloride (PVDC). It is convenient to incorporate foaming in the co-extrusion process to allow simultaneous foaming of the core layer while encapsulating it with a solid or a different density foamed skin. In the presence of a blowing agent, it would be easier to process the specialty polymer and sometimes to reduce the material cost. This technology can be tailored into specific applications such as saving expensive core material or for achieving a more desirable usage of recycled resins in the core layer. However, the expansion in the core layer should be quite limited (under four times), and co-extrusion for foam needs a specially designed combining system for even thickness distribution.

4.7 Areogel

Areogel is not a familiar methodology in polymeric foam industry yet. But it is implied in the cellular technique. Areogel is air in the meshed solid for a very low density structure, some cases under 10 kg/m³. The Areogel process is a staged replacement process. It begins with solid and compatible liquid, and the compatible liquid is replaced by a less compatible liquid.

This step is repeated until a liquid can be replaced by air. Silica gel is a good example [142]. This product is tranparant, and yet with great insulation values. Thermal windows are good applications.

Most polymers dissolve in a solvent at an elevated temperature. The viscous solution can be quenched to a point far below T_g or T_m. Polymer tends to solidify. As enough polymer exists, a gel structure saturated by solvent is formed. Then, slight heating and vacuum can drive the solvent away. A cellular skeleton material is made. Depending on the polymer property, its application includes: sponge for quick sorption and desorption, sound absorption, flow rate control, insulation, astronaut clothing, etc.

This quenching technique is a batch in nature and could make slab and large dimension product for fabricating into a specific geometry for wide usages. When biodegradable polymer is in use, its cellular structure can find applications in hospital, sanitary, and tissue scaffold. It could be an emerging technology for polymeric foams.

5

Relationships Between Foam Structure and Properties

Polymers demonstrate an interesting property range, from as soft as sponge to as hard as rock, from half time expansion to 250 times expansion, etc. The presence of gas basically increases the property dimension of the polymer matrix by expansion to make foam a new product with new properties. The derived mechanical, thermal, acoustic, cushioning properties are unique characteristics for foam, which are important properties for certain areas of our daily life. As a result, foam continues to penetrate into markets for special applications. Some properties as can be conceived are primarily controlled by the amount of gas, and some others by the kind of gas, and the cell size and its distribution. This chapter is focused on the property dependency of the structural parameters.

5.1 Effects of Foam Density

The presence of a blowing agent in the cellular polymer alters its structure so that it could vary from skeleton-struts to integral poly-faced or spherical cells. The nature of the cells comprising the foamed structure, the amount of gas, and its dispersion play critical roles in conditioning the properties of the foam [143].

Density reduction is inherent in polymeric foams. As a result, new properties are associated with the lighter cellular products. Assuming a uniform cell size, the foamed polymer can be conceived as cell-and-polymer composite, as depicted in Figure 5.1. The density reduction can thus be calculated as:

$$\rho_f = (W_g + W_p)/(V_g + V_p) \tag{5.1}$$

where ρ, W, and V represent density, weight, and volume, respectively; the subscripts g and p represent gas and polymer, respectively. Gas is virtually low in weight and high in volume, and polymer is the opposite. At medium

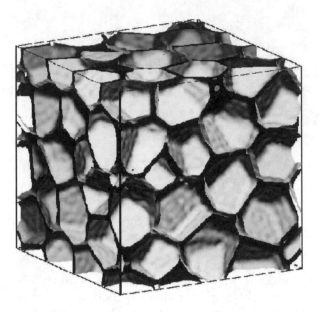

FIGURE 5.1
A typical foam structure consisting of polymer and air.

to high expansion, W_g and V_p become negligible, hence Equation 5.1 reduces to:

$$\rho_f = W_p/V_g \tag{5.2}$$

Gas possesses minimum strength. The more the gas, the less the mechanical modulus. In other words, mechanical property decreases as foam density decreases. Efforts have been dedicated to establishing quantitative correlation between foam density and mechanical modulus. With respect to the modulus, the Cubic Cell model proposed by Gibson and Ashby [144] offers a good correlation between the cell strut and beam strength. As shown in Figure 5.2, the modulus for a pure open-cell foamed structure (only cubic skeleton) has a strong dependence upon the foam density:

$$M_f/M_p = (\rho_f/\rho_p)^2 \tag{5.3}$$

As opposed to open cell, closed cell has not only cell wall buckling, but also cell wall stretch accountable to deformation. The latter, the membrane stretch under stress, has a strong contribution, so that the modulus equation becomes:

$$M_f/M_p = \phi^2(\rho_f/\rho_p)^2 + (1 - \phi)(\rho_f/\rho_p) \tag{5.4}$$

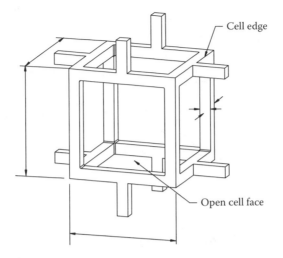

FIGURE 5.2
Strut-Face cubic model.

where ϕ denotes the strut volume fraction. When cell-wall stretch dominates, the above equation becomes:

$$M_f/M_p = (1 - \phi)(\rho_f/\rho_p) \tag{5.5}$$

Since Equation 5.5 is simply a linear density reduction function, as opposed to a square of density reduction valid for open-cell foams, it can be inferred that open cell foams demonstrate a lower modulus than that of corresponding closed cell foams by the expansion ratio. For instance, at 10 times expansion, closed cell foams have a modulus 10 times greater than that of open cell foams. It can thus be understood that spongy open celled foam and rigid closed cell foam can be made out of the same polymer and at a similar density.

It should be noted that ϕ is a function of foaming and expansion for uniform cell wall and strut, ϕ is about $\frac{1}{6}(\rho_f/\rho_p)$. This offers a quick way to assess thermoplastic foam modules.

It is known that gas has a low heat conductivity, which can enhance the insulation as the gas dispersed in the polymer. When foam density decreases, the heat insulation value, known as R-value, increases. It is a common practice for insulated foam in construction to save energy in winter time. Figure 5.3 shows the collected data for different foam at different densities. The trend suggests an increase for decreased density. However, at very low density, another mechanism seems to be effective to cause the decrease.

Gaseous bubbles were found not only to reduce the overall density, but also to function as a "dissipator." The latter virtually slows down both the momentum and heat transfer phenomena. Therefore, energy absorption and insulation become the two most typical applications of polymeric foams. Figure 5.4 demonstrates the effects of foam density change on energy absorption, in which the less the deceleration, the more the absorption.

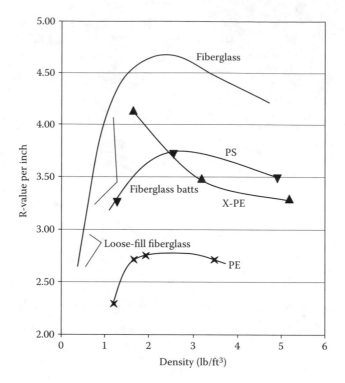

FIGURE 5.3
R-value for fiberglass and foam sheets.

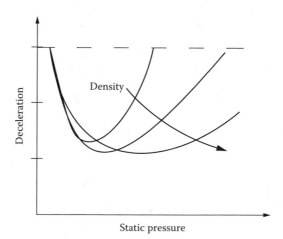

FIGURE 5.4
Deceleration curve at different static loads.

5.2   Open Cell Content and Its Effects

The nature of the foam morphology ranges from pure open cell as skeleton structure to closed cell with complete integral cells. Each structure has specific application benefits. When the target structure is set, the real challenge is how consistently it can be made. It is reasonable to start with the definition. The open cell content is defined as:

$$x = \text{(volume of inter-connected cells)}/\text{(overall gas volume)} \qquad (5.6)$$

It can be experimentally determined by ASTM D-2586-A method.

The open-cell structure plays mixed roles in various applications. On the one hand, open-cell foams are very desirable in gas exchange, absorption, and sound deadening. For instance, a polystyrene tray with controlled open-cell content is necessary to cause capillary flow for fluid retention in the meat tray application [145,146]. On the other hand, open-cell foams are detrimental for keeping the dimensional stability and mechanical properties of the foamed parts. A brief summary is tabulated in Table 5.1.

In polyurethane foaming, the open-cell content is controlled by reaction. However, it does not involve dimensional stability issues even at a high volume expansion (i.e., over 200 fold), because the x-linking generates adequate inter-polymer bonds, which enhance the modulus of the foam so that it can sustain the frame weight.

In thermoplastic foam operations, although they facilitate the gas exchange (mostly hydrocarbon) with air, the open cell content is not desirable for closed cell foams, partly because of reducing the mechanical properties, and partly because the induced dimensional stability concern. Figure 5.5 clearly indicates a reduced mechanical property at increased open-cell contents.

Considering all of the above, the foam mechanical modulus can be derived that for partial open-cell foams, such as foams having holes or weak walls and mixed mode foams, the equation becomes:

TABLE 5.1

Open Cell Effects in Foam Applications

Advantages	Disadvantages
Fast Gas Exchange	Lower Insulation Factor
Low Residual Blowing Agent	Lower Mechanical Properties
Good Sound Deadening	Dimensional Stability Concerns
Enhanced Fluid Absorption	Poor Floatation Property
Good Fluid Absorption and Desorption	Low Energy Absorption
Good Bonding for Dissimilar Coating	Rough Surface

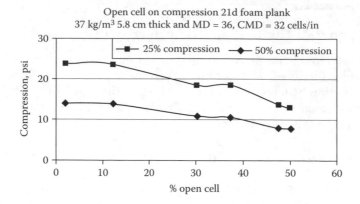

FIGURE 5.5
Variation of PE foam compressive strength at different degrees of open cell.

$$M_f/M_p = (1 - x)\phi^2(\rho_f/\rho_p)^2 + x(1 - \phi)(\rho_f/\rho_p) \tag{5.7}$$

where x denotes the closed-cell fraction. In this context, the variation of the modulus as a function of density with x as a parameter is presented in Figure 5.6.

In foam processing, the open-cell content is primarily dictated by the degree of expansion, but it is also influenced by the processing conditions. Figures 5.7 [147] and 5.8 [118] clearly demonstrate the dependence of the open-cell content on the processing conditions and resultant expansion ratio. For instance, at low volume expansion ratios, the processing window for the control of the open-cell content is relatively wide. However, it narrows as the volume expansion ratio increases, as a higher processing pressure is required to suppress the gas vapor pressure. A respective summary chart is presented in Figure 5.9. It should be noted that the other sets of parameters indicated in this chart include the percentage of blowing agent and the foaming temperature. The chart in Figure 5.8 reveals that low foaming temperatures enhance the melt strength and in general result in foams having a low open-cell content. As opposed to heat induced open cell, the low processing temperature may generate open cell out of lacking stretchability at cell growth. Nonetheless, at high concentrations of blowing agent that are necessary for achieving high volume expansion ratios, the foaming temperature becomes lower due to the plasticizing effects of the blowing agent. An appropriate screw design, capable of keeping a good balance between the mechanical energy input for pumping and mixing, and the thermal energy removal from the barrel, is a key to obtaining the foams with reduced open cell morphology [148,149]. In summary, combination of screw design, processing profile, and resin choice (branched over linear) can greatly enlarge the control window for open cell content.

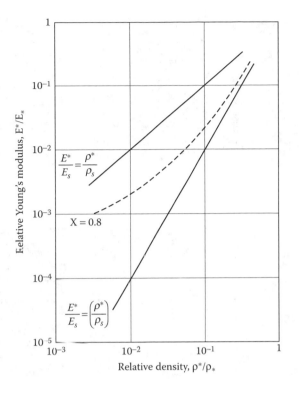

FIGURE 5.6
Relative modulus variation vs relative density, solid curve represents pure open cell, broken line represents closed cell and x is the open cell fraction.

5.3 Key Properties: Energy Absorption, Insulation, Sound Absorption, and Mechanical Properties

The presence of gaseous bubbles (cells) in foamed polymers generates unique physical and mechanical properties in comparison with the base unfoamed polymer. The lower density grants buoyancy to foamed-polymer parts, whereas the softness, energy absorption, and thermal insulation capability open possibilities for a great variety of applications such as floatation, construction, automotive, transportation, sports, medical, furniture, and so on. Their performances are closely associated with material properties. The relationship becomes extremely important when new foamed products have been, or are being, developed for more intensively performance-oriented applications.

According to Figure 5.10, the closed cell structure appears to be more capable of resisting external disturbances. It demonstrates another distinct feature of foamed polymers, i.e., cushioning capability. When the polymer

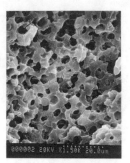

(a)   (b)   (c)
$T_c = 170°C\ T_n = 175°C$ $T_c = 170°C\ T_n = 135°C$ $T_c = 170°C\ T_n = 110°C$

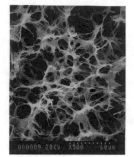
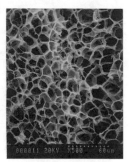

(d)   (e)   (f)
$T_c = 150°C\ T_n = 175°C$ $T_c = 150°C\ T_n = 135°C$ $T_c = 150°C\ T_n = 110°C$

(g)   (h)   (i)
$T_c = 120°C\ T_n = 175°C$ $T_c = 120°C\ T_n = 135°C$ $T_c = 120°C\ T_n = 110°C$

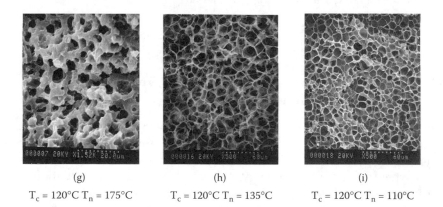

FIGURE 5.7
Processing temperature effects on foam structure. (C.B. Park et al., Low-Density, Microcellular Foam Processing in Extrusion Using CO_2, *Polym. Eng. Sci.*, 38, 11, 1998. With permission.)

is elastic enough, energy absorption by deforming the cell to dissipate the impact can be instituted for specific applications such as seat cushions, impact protection, and shock absorption.

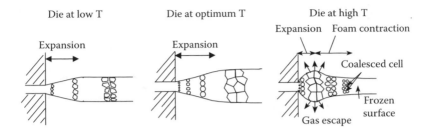

FIGURE 5.8
Die temperature effects on foaming PP. (H.E. Naguib, Fundamental Foaming Mechanisms Governing Volume Expansion of Extruded PP Foams, Foams 2002, *Soc. Plas. Eng.*, 2002. With permission.)

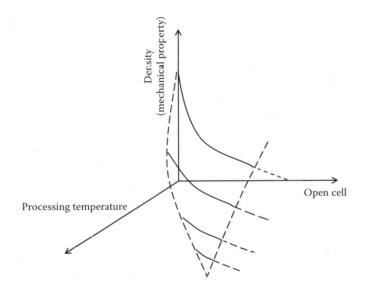

FIGURE 5.9
General processing morphology density (property) chart for thermoplastic foams, solid line represents operational window.

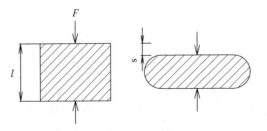

FIGURE 5.10
Closed cell under compression; wall stretch and gas pressure.

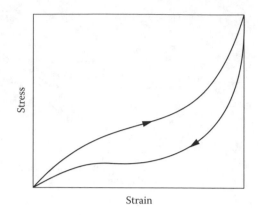

FIGURE 5.11
Stress-strain curve for flexible PU and PE foam; < loading and ; unloading direction.

In a typical stress-strain curve, the area under the curve represents the energy required for the strain, as depicted in Figure 5.11. It also shows the desorption curve that often is not exactly the same as the absorption curve owing to the visco-elastic nature of polymer. The hysterisis is the difference in area. A perfectly elastic material is supposed to have a very small hysterisis, and is thus good for repeated usage. Polyurethane foam for seat cushion and x-linked polyolefin foam for shoe sole are good examples.

For a closed cell structure, if elastic cell walls and a zero Poisson constant are assumed, the compressive strength, τ, can be expressed as:

$$\tau = \tau_o + (1 - x)P_o\varepsilon/(1 - \varepsilon - \rho_f/\rho_p) \qquad (5.8)$$

where P_o and ε denote the cell pressure and strain, respectively, whereas the subscript o means the initial state. When the open-cell content varies owing to cell wall rupture, the deviation from Equation 5.8 increases. However, for a non-zero Poisson constant, the expression can be found in the reference literature [150].

With respect to the impact absorption, a similar implication regarding the deceleration factor at the simulated object drop would be valid. It is defined as:

$$G' = 2(\text{Impact Energy} - \text{Absorbed Energy})/(mv^2) \qquad (5.9)$$

where m and v denote the mass of the object and the instantaneous velocity, respectively. Since a low object deceleration indicates a high energy absorption by the foam, it can be inferred that the external impact can cause internal cell structure damage, i.e., cell rupture or cell weakening. The deceleration factor increases as repeated tests are conducted. At any event, the repeated low deceleration is ideal for packaging and transportation protection. Typical deceleration curves are presented in Figure 5.12. It is important to note that

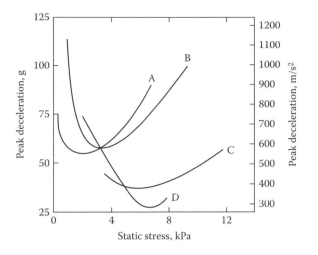

FIGURE 5.12
Dynamic cushioning curves for different polymeric foams, A: PU, B: LDPE, C: PS bead, D: PS (adapted from [151]).

the lowest point indicates the best energy absorption, which is very useful for the selection of foamed polymers and for designing packaging applications [151].

For a long number of years, polyurethane foams have been used for insulating refrigerator doors and, along with polystyrene foams, for the manufacture of insulation boards as construction materials. The insulation factor is an energy conservation parameter. The overall heat transfer factor (k) is determined by the composite structure. For instance, a refrigerator door comprises a frame and a foamed insert, so that the composite k factor is:

$$k = k_m + k_f = k_m + k_p + k_g \qquad (5.10)$$

where the subscripts, m, f, p, and g, represent metal, foam, polymer, and gas, respectively. Usually, the gas contribution ranges from generally over 50% to as high as 75% [36]. The contribution of each component to the overall thermal conductivity of the PU foam made with CFC-11 is illustrated as a function of the foam density in Figure 5.13 [152].

Nowadays, the search for alternative blowing agents becomes a critical environmental and energy dilemma for rigid polyurethane foam producers. Since halogenated hydrocarbons are being phased out, the search for environmentally safe and low-conductivity blowing agents becomes a very crucial research subject for the appliance industry. Recently, heavy HFCs have been a priority research subject. But chlorine has a much more desired stability over fluorine, which becomes a major challenge in the replacement search.

Sound deadening is another interesting area for applying polymeric foams. This is so because when the sound waves contact the surface of a foamed

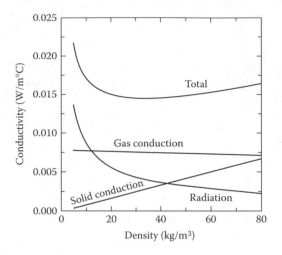

FIGURE 5.13
Thermal conductivity and the elements at different density. [152] with permission.

product, the typical surface (micro-valley and hill) cause additional bounces thereby dampening out the intensity of the wave before reflection. It is noted that audible sound has a wide spectrum, and it is difficult to find a foam good for the whole range. However, when open-cell structures are present, the deadening effect of the foam is significantly augmented. But, since open-cell foams demonstrate energy absorption capability loss and mechanical property reduction, a trade-off has to be made. Nevertheless, since polymeric foams could be made to have dual properties; i.e., both sound and energy absorption, they represent an excellent material for ceiling and floor under-layment. A typical sound absorption curve is demonstrated in Figure 5.14. Lately, another membrane vibration mechanism was proposed to describe the dampening out phenomena in big cell foam [153]. As opposed to the open cell structure, a tiny flaw in the cell wall may damage the attenuation capability.

Other unique properties, such as antistatic, flame retardancy, ultra violet resistance, vapor corrosive inhibition, etc., could be generally obtained by adding proper additives, which can be found in a plastic handbook [154]. However, it should be pointed out that some additives, especially the inorganic metal compound with a very high melting characteristics, can cause processing and nucleation concerns and limit the processing para-meters within quite a narrow foaming window. Also, when the dosage of the additives is relatively high, it may affect the cell wall integrity. Since most additives are not compatible with polymers, the concomitant phase separation at high expansion ratios will certainly weaken the properties of the foam.

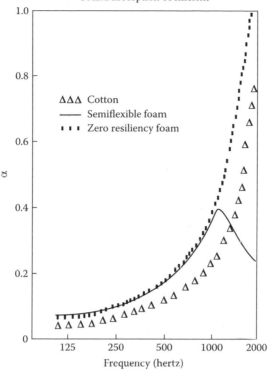

FIGURE 5.14

Sound absorption chart. (R. Herrington, Dow Polyurethanes: Flexible Forms, Dow Chemical, 1997. With permission.)

5.4 Cell Size Distribution and Its Effects

At a given expansion ratio, cell size may cause significant effects in mass and heat transfer due to the cell wall thickness and convection in the cell. For instance, cell size effects on thermal flow are depicted in Figure 5.15. It clearly indicates that the thermal properties of the foam are strongly affected by the cell size and the way the cells are dispersed throughout the occupied volume. It should be noted here that at medium and low expansion ratios, it is reasonable to assume a spherical cell shape and a uniform cell size, whereas for high expansion ratios, the cell size tends to show a distribution owing to the nonsynchronized nucleation and cell coalescence.

Cell nucleation is an energy-controlled process in which stable nuclei gather an energy large enough to overcome the surrounding confinement to develop bubbles into the unstable growth state governed by transport and thermodynamic properties. At impulse nucleation and low-degree

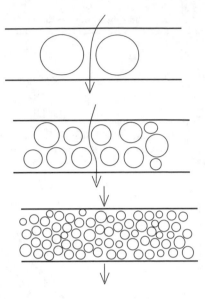

FIGURE 5.15
Heat flux for different cell size foam.

expansion, the cell size should be uniform. However, in reality, an even energy distribution is difficult to control, and, as a result, "staged" nucleation often occurs. That means, instead of being instantaneous, the nucleation begins at one point and continues on [155]. A typical example is that gas evolution out of a reaction is not uniform from place to place. Although interfacial expansion from the primary birth creates more surface area to facilitate diffusion, local gas generation can still generate a favorable condition for successive birth. Sometimes, secondary nucleation, such as induced by cell wall stretch [156], becomes non-negligible. In addition, cell coalescence can occur depending upon the prevailing conditions in the surrounding, which contributes to the creation of a non-uniform cell size distribution. At high expansion ratios and/or at high cell densities, the cell size distribution virtually exists.

Figure 5.16 shows the cell size distribution for extruded polystyrene foam, in which a normal distribution appears. Since the properties of a foamed product are primarily determined by the amount of polymer and the integrity of the cell wall in a low-density foam, the property changes across the cell wall thickness can be approximated by a linear model. Then, the cell wall thickness will not affect the foam properties as long as the foam density and cell integrity are similar. In this context, Figure 5.17 depicts the marginal effects of the cell size distribution on foam properties. However, when the foam expansion ratio increases, it is likely to exceed the threshold in generating a weak cell wall or a broken cell wall, which will greatly reduce its mechanical properties [157].

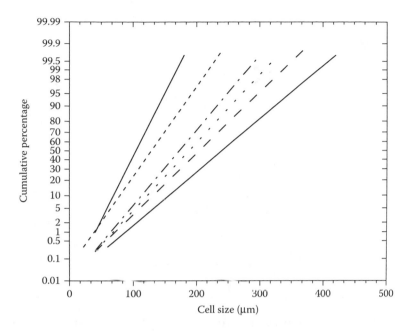

FIGURE 5.16
Cell size distribution for PS foam; expansion increases toward left (25 to 75 kg/m³).

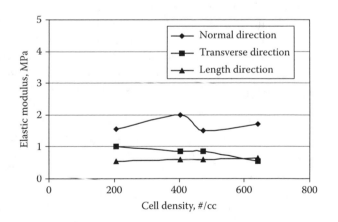

FIGURE 5.17
PE foam compressive strength vs cell size.

As mentioned in Chapter 3, when the expansion exceeds four times, or 75% gaseous voids, the contact with the neighboring bubble becomes inevitable [115]. Consequently, spherically shaped and uniformly sized cells cannot co-exist in that case, and cell distortion will occur (see Figure 3.19). However, as shown in Figure 5.18, a secondary nucleation to fill the inter-bubble space can ideally reduce the density without sacrificing the spherical

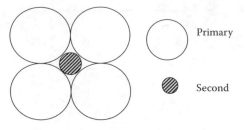

FIGURE 5.18
Body-centered packing with secondary bubble.

shape. It adds to the degree of cell-size distribution. It is thus conceivable that dual or triple nucleation can maintain cell integrity at a very low density.

5.5   Dimensional Stability

In general, the foaming process involves mixing and dispersion of a blowing agent within the polymer as well as a subsequent dynamic expansion to create the cellular structure. It inherently involves heating-and-cooling and pressure-and-depressurization. After the foamed product is made, naturally mutual diffusion and cooling normally occur to allow the exchange of the blowing agent with air and stabilization to room temperature for further fabrication or direct application. During aging, the low modulus polyethylene foam tends to possess the collapse-then-recover aging characteristics as illustrated in Figure 5.19. Depending on the thickness, it may take months to recover, which evidently became a production concern.

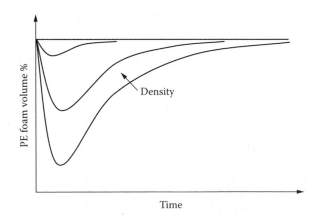

FIGURE 5.19
Polyolefin foam dimensional variation during aging.

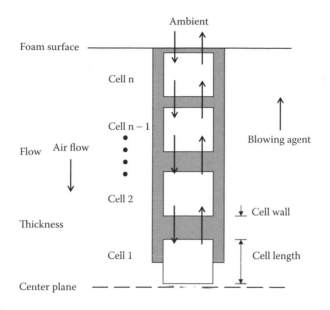

FIGURE 5.20
Cubic Structure Model.

A viscoealstic model, as illustrated in Figure 3.25, was presented to describe the cell wall and cell pressure variation with respect to the dimensional variation of the cell during aging [84,158]. Foam is simulated as a cubic structure as shown in Figure 5.20. The permeation difference between the air and the blowing agent can cause significant pressure losses in the cell, which can, in turn, generate foam thickness losses and result in a recovery curve as shown in Figure 5.21. This occurrence is frequently taking place in over 30-fold expanded polyethylene foams.

In general, depending upon the blowing agent and air permeability ratio and the dimension of the product, it may take hours, days, or even years for the foam thickness to recover to its original dimension. It should be also noted that since most physical blowing agents are of hydrocarbon structure, their flammable nature makes the residual blowing agent in the foam highly undesirable. It thus clearly becomes a technical challenge to successfully maintain the dimensional stability of the foam without sacrificing its safety status.

There are both chemical and mechanical remedies that can eliminate concerns related to foam aging. One way of dealing with these issues includes adding additives into the formulation to regulate the permeation ratio and thereby keep the dimensional stability of the foam under control. Another way is based on blending resins to enhance the cell wall strength and thereby minimize or avoid the stability problem [159].

Although the dimensional losses of the foam could be prevented with good x-linking, PU foams in fact, can be made to have a high degree of open cell content to allow fast permeation. In addition, it is suggested to adopt a

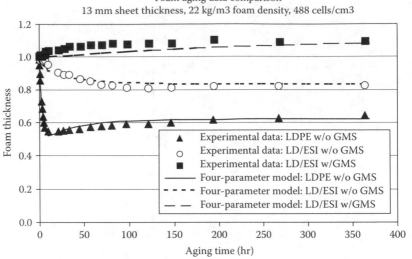

⌐IGURE 5.21
PE foam aging curve; experiments and simulations. (C.T. Yang et al., Dimensional Stability of LDPE Foams: Modeling and Experiments, 59th *Ann. Tech. Conf. Soc. Plas. Eng.*, 2001. With permission.)

long curing period so that the exchange of the blowing agent with air becomes completed before forming the product into a roll stock [160,161]. Commonly, by implementing tiny holes on the surface of the foam after solidification has been achieved the air/blowing gas exchange can be further facilitated [162,163].

It has been also noted that when over-expansion of foam occurs due to too much blowing agent added, or due to too high foaming temperatures, relatively immediate shrinkage takes place. This anomaly can be avoided by correcting the formulation and/or by adjusting the foaming temperature. However, inadequate mixing and non-uniform cooling can cause uneven expansion and thus non-uniform cell size distribution throughout the cellular structure, which, in turn, can create subsequent dimensional stability concerns. These concerns are mostly related to processing and processor issues.

The residual thermal stresses, especially with respect to bulky foams obtained by injection molding or extrusion, represent another interesting area for research. The radial stretch induced stress is inevitable for foam expansion, but since it is in the molecular domain, it does not represent a major concern in real operations. However, since polymers are poor thermal conductors, and since foamed polymers are even worse, the increase of the relaxation time during cooling to the temperature of the surrounding can easily exceed the cooling time. This indicates that thermal stresses are being built in the product during foaming, which is especially true for the injection molding process during which the foaming occurs under the mold pressure.

At the time of mold release, if the foam is not cold enough to retain the geometry, immediate distortion may occur. Sometimes, although the quality of the obtained foam may initially appear satisfactory, in the case of bulky foam products, the thermal treatment in the subsequent fabrication could cause heat-induced uneven thermal stress relaxation (i.e., bowing and bulging). Often, mold design modifications, flow and cooling cycle are more practical approaches to offset these stresses.

Another perspective of stability is the increase in thickness during thermoforming. Although soluble pentane or butane is preferred in making a polystyrene sheet, which is, in turn, formed into tray, at the heating stage of the thermoforming operation, the soluble hydrocarbon tends to diffuse into the cell and expand in response to heating so that tray thickness could expand more than 100%. However, when an insoluble blowing agent, such as carbon dioxide, is applied, the heat-induced expansion is much lower. This indicates that the increase in thickness in this case is solubility and residual blowing agent governed matter. Although it is virtually integrated into the final thickness and cell size for PS and PP foam thermoforming, the selection of the blowing agent and its residual level at different aging times are the main concerns for successful thermoforming.

Table 5.2 shows typical dimensional stability concerns and the possible remedies. Since most solutions have some side effects, especially if over-exercised, for achieving consistent product quality, it is highly recommended to maintain a good balance between the formulation and the processing/foaming control. Using a statistical approach may prove very useful in finding the optimal processing window for manufacturing.

TABLE 5.2

Dimensional Stability Issues, Causes, Effects and Remedies

Phenomena	Causes	Remedies
Immediate shrinkage	Over-expansion (Too high open cell content) Too much B/A, Too hot in foaming	Formula, Foaming and processing
Gradual shrinkage and slow recovery	Fast B/A permeation B/A condensation	Additives Surface coating Hot aging
Distortion	Thermal stress Foaming under pressure	Storage for relaxation Uniform thermal and pressure at foaming
Expansion	Slow B/A permeation Soluble B/A	Right B/A for thickness increase

5.6   Residual Blowing Agent

Most of the blowing agent expanded foam has an aging process during which air replaces the blowing agent. It can last from seconds to years. Often, blowing agent and air are quite different in transport, physical and thermal properties. The general impacts are as below:

1. properties: insulation property and mechanical property
2. safety in fabrication or storage for flammable blowing agent
3. post-handling distortion owing to uneven permeation
4. environmental emission status

The residual gas and air concentration within the polymeric foam is a function of time with opposite trends, as depicted in Figure 5.22. The gas possesses different heat conduction and compressibility, which can further cause insulating effects and/or compression variation as gas concentrations vary. Most blowing gases are better insulators than air. This means that the heat transfer ability of the foam increases as more and more air replaces the blowing gas in the foam, as illustrated in Figure 5.23 [122]. It is also important to note that the mechanical properties of the cells decrease as more blowing agent is dissolved in the polymer. Figure 5.24 shows a significant decrease as blowing agent dissolution increases [164]. A proper choice of the blowing agent and an appropriate "aging" for property maturity is necessary practice in foam manufacturing.

When the blowing agents that represent environmental concerns are employed in the manufacturing process, certain regulations regarding emission or safety have to be met. "Recovery for reuse" is one way of reducing safety concerns related to expensive blowing agents, such as

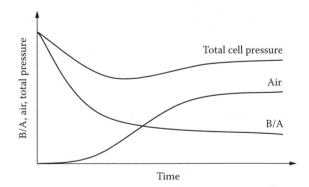

FIGURE 5.22
Pressure variation in PE foam aging.

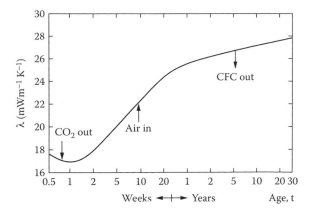

FIGURE 5.23
Effects of residual blowing agents in PU foam. (C.J. Hoogendoorn, Thermal Aging, in *Low Density Cellular Plastics*, Hilyard, N.C. and Cunningham, A., Eds., Chapman & Hall, London, 1994. With permission.)

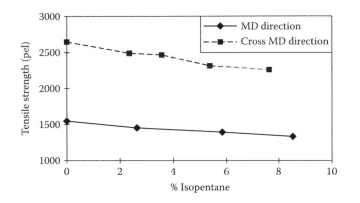

FIGURE 5.24
Tensile strength reduction at increased isopentane in LDPE film. (S.T. Lee and K. Lee, Solubility of Simple Alkanes in Polyethylene and Its Effects in Low Density Foam Extrusion, Foam Conference, RAPRA, Frankfurt, Germany, 2001. With permission.)

hydrofluorocarbon (HFC). Du Pont proposed a recovery unit with active carbon to adsorb halogenated blowing agent, which was then stripped by steam for recovery [165]. On the other hand, for inexpensive hydrocarbons, the main concern is the residual blowing agent induced safety concern. Since the properties and availability of hydrocarbons make them more economical for foaming than carbon dioxide (because the foaming with carbon dioxide results in higher density foams), attention is more focused on the handling and safety.

Common hydrocarbons have explosive ranges as tabulated in Table 5.3. Therefore, although hydrocarbons are very compatible with general thermoplastic polymers and are popular in making foams, the accumulation of the

TABLE 5.3

Explosive Limits for Some Common Blowing
Agents

Blowing Agent	Lower Explosive Limit	Higher Explosive Limit
CFC-12	none	none
CFC-11	none	none
HCFC-22	none	none
HCFC-141b	1.3	16.0
HCFC-142b	6.7	14.9
HFC-134a	none	none
HFC-152a	3.9	16.9
HFC-245fa	none	none
HFC-365mfc	3.5	9.0
Ethane	3.0	12.5
Propane	2.2	9.5
n-Butane	1.8	8.4
i-Butane	1.9	8.5
n-Pentane	1.4	8.3
CO_2	none	none

permeated hydrocarbons in a non-vented space can certainly cause safety concerns. Therefore, methods that offer accelerated replacement of the hydrocarbons with air are necessary to make the foamed product absolutely safe. Commonly, thermoplastic foam producers implement methods that include blending the blowing gas with inert gases, festooned processing (i.e., longer traveling for foam sheet before winding up into roll stock), and drilling artificial holes.

When the foam volume expansion ratio exceeds 40 times, it is not uncommon to have over 10% gas in the fresh product. If the dimension is large enough, often secondary fabrication operations are needed for achieving a given geometry for specific applications, so that in these cases it is relatively easy to develop fresh surfaces. However, when the surfaces are not even, because of uneven diffusion, temporary or permanent dimensional distortions can occur, depending upon the permeation time and relaxation time ratio. This is evidently a quality issue. In other words, keeping a low level of residual blowing agent is a necessary exercise for safe and sound handling of bulky foams made with hydrocarbon-based blowing agent components.

5.7 X-Linked Content

Polyethylene (PE) can be classified by its branching content and length. In general, it could have a linear structure, or short and long side chain distribution. In this context, various types of PE grades can be distinguished, such

(a) Peroxide
    a. $ROOR + heat\ or\ hv \rightarrow 2RO\bullet$
    b. $RO\bullet + PH \rightarrow P\bullet + ROH$
    c. $P\bullet + P'\bullet \rightarrow P - P'$

(b) Irradiation; electron beam, Co60
    a. $PH + eV \rightarrow P\bullet + H\bullet$
    b. $P\bullet + P'\bullet \rightarrow P - P'$

FIGURE 5.25
X-linked mechanism for PE foam.

as high-density polyethylene (HDPE), linear low-density polyethylene (LLDPE), and low-density polyethylene (LDPE). The long-chain branching in LDPE develops unique strain-hardening characteristics, which are necessary for good foaming. As illustrated in Figure 5.25, further chain extensions or x-linking could be made by introducing a free particle generator (i.e., peroxide) to create stable secondary free particles by attacking the hydrogen atom to generate bonds. Thereby, the melt strength of the resin can be enhanced, which is very helpful in foaming [166,167]. Figure 5.26 demonstrates the x-linking effects in foaming and the optimal point that exists for foam expansion [168]. It should be noted that when the melt strength increases simultaneously with the increase of the x-linking content, the nucleation can be conveniently controlled so that obtaining a fine- celled structure

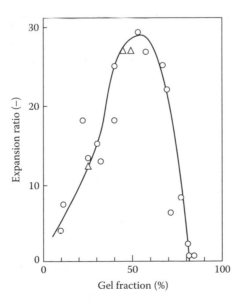

FIGURE 5.26
Expansion ratio of reactive polyolefin. Circle represents Reactive Polyolefin; triangle represents 50% reactive polyolefin + polypropylene. (S. Gotoh et al., Reactive Polyolefin, *SPE ANTEC*, 1986. With permission.)

becomes feasible. Namely, it virtually becomes another interesting feature of x-linked PE foams.

Polyethylene is difficult to dissolve in organic solvents at room temperature, but it tends to dissolve in xylene at an elevated temperature. It is important to note here that after the x-linking is completed, the cross-linked gel is not dissolvable in the hot xylene. This, however, becomes a reliable method to check out the x-linking percentage by measuring the portion of undissolved PE in the hot xylene bath. The level of x-linking in PE foam evidently affects its properties and applications.

With respect to the properties, the more x-linking, the more chemical bonds are formed for generating greater strength. Cross-linked PE grades, however, demonstrate more thermal than mechanical improvements over respective non-x-linked ones. Basically, the x-linking forms networks between polymer backbones, which certainly reduces melt flow capability. In fact, at temperatures much greater than its melting point, it is definitely in molten state, but not ready to flow. It thus widens the thermal stability window that renders the resin suitable for various thermoforming applications [169]. The cross-linking tends to generate more extension and links which results in long relaxation times. It tends to reduce crystallinity, but greatly enhances melt viscosity. As a result, obtaining a foam with a fine celled structure becomes feasible, which, due to the thin cell wall, gives a very soft touch feel. This is actually a unique feature of x-linked PE foams.

It should be pointed out that x-linked foams possess enhanced chemical resistance, thermal, and sometimes mechanical properties. However, the drawback is its aggravated recycling and reuse because the extra bonds make it extremely difficult in the reprocessing attempts. Efforts are required to break down the x-linked structure and thereby improve its ability for being reused and/or reprocessed.

6

General Polymeric Foams and Their Applications

6.1 Flexible and Rigid Polyurethane (PU) Foams

Polyurethane foams, flexible or rigid in nature, were developed as early as in the 1930s. Since then, they have been established in numerous extensive as well as intensive applications, especially in the post World War II period. The main feature of a PU foam is its capability to deliver a wide range of densities, cell structures, foam morphologies, and rigidity, controlled solely by formulation adjustments. The key markets and applications for PU foams are listed in Table 6.1.

After 60 years' practice, PU foam is quite predictable in performance. It is known for its surface feel, strength, and durability. In general, flexible PU foams made with shorter polyol and less functional groups demonstrate excellent elastic and deformation/recovery characteristics suitable for seat cushioning, flexible hoses, packaging, sports, recreation, etc. In contrast, rigid PU foams possess not only rigidity, but also insulation capabilities. Therefore, rigid PU foams became a popular material for construction, appliance, recreation, sports, and automotive applications. It should be noted that the

TABLE 6.1

General Applications for PU Foam

Items	Markets	Applications
Flexible PU	Transportation	Seating, pads, liners, dampening, carpet backing, filters, flooring, armrests, trim
	Furniture	Bedding, padding, flooring
	Recreation	Sport mats, toys, helmet liner, chest protection
	Packaging	Electronic, computer, china, equipments
Rigid PU	Construction	Insulation, flooring, siding
	Appliance	Refrigerator frame, Door, Dishwasher door
Semi-Rigid	Automotive	Dash panel, liner, viser
	Footwear	Soles

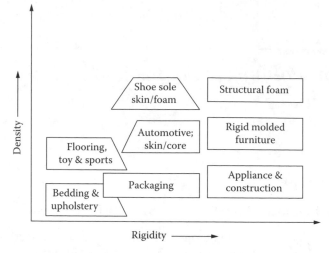

FIGURE 6.1
Application and density chart of PU foam.

rigidity is independent of foam density. Either type can foam into a wide density range. The application and density chart is illustrated in Figure 6.1.

Owing to their versatile material performance PU foams are used in a great number of applications. The majority of applications are dictated by the PU structure, but some are determined by the type of blowing agent. Since environmental regulations became a serious issue, it has been difficult to find an adequate replacement for CFC-11, which is ideal for foaming PU: stable, low toxic, easy to handle, and provides high insulation. Therefore, the research focused on finding an alternative blowing agent becomes a top priority for CFC and HCFC blown PU foam. The current focus is on HFC and HC blowing agents [40,41]. However, either one may require adjustment in application codes to ensure safe and friendly usage.

In the area of furniture seating, flexible PU foam has become the choice product not only because of the economics offered by large-scale productions, but because of the performance characteristics so easy to be adjusted to fit the requirements. The latter evidently helped the former to make it quite popular in seating applications. Automotive seating is a typical example, as illustrated in Figure 6.2.

Another application for the flexible PU foam is the floorunderlayment (or carpet) industry. The foaming mixture was dispersed over the back of a carpet, which moved through a fixed blade to allow even foaming thickness. Its advantages are shock, heat, sound insulation, which were recognized in the sixties. PU foam in the flooring of residences and offices is quite common now.

As shown in Figure 6.3, PU foam is excellent in the packaging industry. Its slow foaming and filling fit nicely in shipping objects via carton box. Incorporating into the continuous line, it has been widely used in shipping products such as books, electronics, automotive parts, etc. Its vibration and

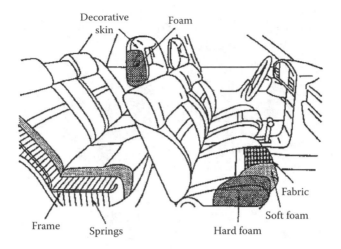

FIGURE 6.2
Automotive seat application for flexible PU foam. (R. Herrington, Dow Polyurethanes: Flexible Forms, Dow Chemical, 1997. With permission.)

Foam-in place:
A simple cushioning or blocking and bracing process for a variety of items with changing shapes and sizes.

| 1. Instapak® foam is dispensed into a carton lined with high-strength Instamate® film. | 2. The Instamate® film is folded over and the product is placed on the rising foam. | 3. A second sheet of Instamate® film is placed over the product and more Instapak® foam is dispensed. | 4. Your customer receives your product undamaged. |

FIGURE 6.3
Application of PU foam in packaging industry. (Courtesy of Sealed Air Corp.)

attenuating characteristics have proved fitting in shipping and transportation. Sealed Air Corporation had a significant success under the trade name Instapak® since the 1970s. It can also be molded like convoluted surface, and two sets of that to form a interlocking for packaging as illustrated in Figure 6.4.

Semirigid PU foam is made with increased x-linking and polyol copolymer to improve its hardness and energy dissipation properties. It is perceived as energy absorption foam, but after drastic impact, it tends to deform permanently. These foams are used in the automotive safety design: interior of bumpers and door panels to dissipate impact.

FIGURE 6.4
Convoluted surfaces on the foam for interlocking.

Rigid PU foams demonstrate high mechanical modulus, thermal insulation and stability. It can be filled into simple panel and complex shape for a variety of applications. Refrigerator doors are a good example, as shown in Figure 6.5. Other applications are monitor cases, rigid furniture, and roof construction. Sometimes post foaming fabrication is necessary to cut the slab into prescribed dimension for end users.

The molding and fabrication capability can easily lend PU foam into toy, sports, floating, casting, carving, sculpting, and taxidermy applications. Considering a wide range of density and structure, and combining with other materials for surfacing, and reinforcement for further exotic usages, it is

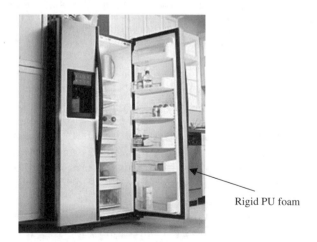

Rigid PU foam

FIGURE 6.5
Application of rigid PU foam in a refrigerator door.

simply amazing to see such as wide and deep usages nowadays. No wonder, it is the primary products among the polymeric foams.

6.2 Polystyrene Foams: Sheet and Slab

Polystyrene (PS) foams are one of the most developed kind of plastic foam. The low glass transition temperature of PS along with the amorphous structure and its capability for being processed in both batch and continuous foaming excited early researchers. Because of its low density and rigid nature, PS foams found a wide variety of applications [170]. In the 1990s, the annual production reached billions pounds, reaching the highest ratio between foam and unfoamed thermoplastic polymers.

Because of the amorphous structure of PS, its foams show favorable stretch-induced hardening [171], which became the benchmark characteristics for foam for over twenty years (1940–1965). Extruded PS foams found applications in floatation in the 1940s. Besides, the glass transition temperature, T_g, of PS, which is around 105°C, lends itself to soaking and foaming with the aid of steam into molded bead foam. General halogenated hydrocarbon, hydrocarbon blowing agent, and even carbon dioxide have good solubility in polystyrene. Therefore, PS foams became the first commercial thermoplastic foam product.

In addition, PS has a higher flexural modulus at room temperature in comparison with other thermoplastics. This indicates that at the same density PS foams demonstrate a better rigidity, or at the same flexural strength, less PS material is needed with a lower density. Such advantages in material properties place other polymers out of competition against PS in the thermoforming area. Moreover, the rigidity advantage, coupled with a fine-celled structure, is ideal for insulation applications [32]. The other application is the molded bead and loose peanuts in packaging. After the PS beads soak in the blowing agent, they can freely expand into loose peanuts or form and fuse in a mold. In brief, the major markets for PS foam are food, insulation, and packaging.

In the food industry, extruded PS sheet can be thermoformed into a shallow to medium deep tray. Its modulus, thermo strength, and friendliness in food contact made it a benchmark for the market. Since PS has a relatively poor barrier property, and foam makes it worse, often it is necessary to laminate coextruded barrier film to PS foam to enhance the barrier property to improve the freshness of the meat object.

In thermoforming, a simple criterion for cylindrical mold can be derived from the ratio of formed area over the original area. The ratio is:

$$Ra = 4(H/d) + 1$$

TABLE 6.2

Selected R-Value from Different Foams at 1" Thick

Foam Products	Density, lb/ft³	Density Kg/m³	R-Value h-ft²-°F/Btu	Conductivity W/(m° C)
PE Foam Slab	2.2	35	2.8	0.051
X-PE Foam	1.9	30	4	0.036
PS Foam Slab	2	32	5	0.029
PS Foam Bead	1.5	24	4.2	0.035
PP Foam Slab	1.2	19	3.74	0.039
PU Foam	1.5	24	6.25	0.023
PET Foam	8.8	140	3.73	0.014

where, H and d represent height and diameter of the mold. In the light of a given modulus, the ratio can help us understand how much weight it can hold after formed.

The PS tray is in the meat and poultry department of almost every supermarket nowadays. An absorbent pad is placed on top of the tray to absorb the remaining liquid to prevent bacteria propagation. In certain cases, open celled PS is used as the main mechanism to absorb and retain the juice through capillary effects. However, PS has a T_g slightly over 100°C, which becomes a major drawback in packaging instant food for microwave heating.

Combining insulation, rigidity, moldability, and food contact property, the hot drink cup became another huge success for molded bead PS foam. Also, it was used in the clam shell container for fast food. But due to volume concerns in landfills, McDonald's switched to corrugated paper in the late eighties. Future usage depends heavily on regulatory development.

Another section for extruded PS foam is the insulation board. Builders, architects, and residents have used it as panel board for more than 50 years. The thermal insulation capability for a material is expressed as R-value, which is the inverse of thermoconductivity at 1" thick. Industry rates the insulation with R-value. That means to match the R-value with thickness. The better R-value material, the thinner to match the requirement. The selected R-value for the well established products are listed in Table 6.2. Figure 6.6 shows the cell size effects on conductivity. Fortunately, making fine celled PS is not a difficult job. But its moisture resistance is not as good as PU foam. When moisture is picked up, the R-value drops.

One more major application is in packaging. Although PS itself is quite rigid, after high expansion the cell wall is thin enough to feel soft and become suitable in packaging. PS foam can be made from two different methods; one is the molded bead, which can be molded into corner shapes to protect heavy objects such as appliances in transportation. The other one is the loose fill peanuts to fill the void in the carton box. Figure 6.7 demonstrates the applications. We know the loose fill tends to fly around. If the user is not careful enough, unpacking may become a mess.

Thin PS sheet can be thermoformed into multiple simple shapes for simple objects in transportation. Utilities and cosmetic boxes are good examples.

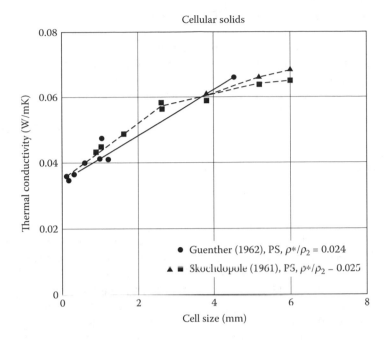

FIGURE 6.6
Effect of cell size on conductivity.

FIGURE 6.7
PS foam peanuts in packaging applications.

But a packaging concern is that after repeated rubbage against the surface of extruded sheet may leave marks on the polished surface.

After over 50 years commercial usage, PS foam has penetrated into our daily lives. It became indispensable to our living in the areas of food container and house insulation. The known drawbacks are that rigid PS foams are prone to cell wall damage and show relatively low recovery after being exposed to high intensity impact forces, which significantly reduces their suitability for repetitive usage. The lack of soft feel of touch also limits the application spectrum of PS foams. The major variable in the future PS foam application seems to be the upcoming discoveries in its effects on environment and health.

6.3   Polyolefin Foams: PE, PP and Blends

The large consumption of polyolefins, ranked as number one in the world, continues to draw deep developmental efforts to improve resin manufacturing, which becomes a phenomenal support to the foam industry. Polyolefin foams can consist of polyethylene, x-linked polyethylene, and polypropylene. The discovery of halogenated hydrocarbons fostered the foaming technology which enabled the production of low-to-medium density foams and thin-to-thick boards for a wide variety of applications. Since the 1960s, polyolefin foams have been manufactured using continuous extrusion-based processes and physical blowing agents. Since the 1970s, both molded bead batch processes and oven heat processes have been implemented as well.

Being distinct from polystyrene foams, polyethylene foams are generally soft and resilient, and possess good chemical resistance [106]. Low-density foams with over twenty-fold volume expansion ratios are normally made of low density polyethylene (LDPE) which is known for its strain-hardening characteristics. Foam extrusion is a common method of making low density PE foams by blending CBA or PBA in the extruder. PE foams basically follow the growth pattern of PS foams. PE and PP molded bead foams became commercial products in the 1980s, especially for achieving special shapes that cannot be manufactured using general extrusion equipment. It simply combines foam and molded part in the same process. PE foams are most suitable for packaging, flotation, seat cushioning, and recreation applications.

In general, PE foam has thin sheets (under $1/2$"), laminated slab, and thick board (over 1"). Thin foam is primarily for surface protection, an important part of packaging, which other polymeric foams cannot fit economically or performance wise. A typical illustration is the interleave between china wares as illustrated in Figure 6.8. When thickness built up via lamination or slab structure, the flotation benefits were recognized. Life jackets and surf boards involve thick PE foam and have been on the market since the 1980s. Slab foam possesses special cushioning performance even after repeated impact, which could cause clear damage to some rigid foams, including PS foam. Its cushion and pliable nature helped post fabrication design to allow special configuration for unique applications. In Figure 6.9, pulling up the

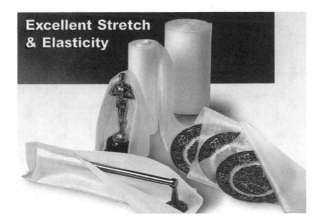

FIGURE 6.8
Surface protection with a thin PE foam. (Courtesy of Sealed Air Corp.)

designed pieces, lap top computers can fit snugly in the space of an ideal packaging design. This simply opened a gate for solution design, which is very unique for PE foam. When density of the slab foam increases, its capability to support also increases. Applications have been found in relatively heavy duty transportation. The military is a good example of heavy density PE slab foam application.

Coating with cohesive makes a self-seal pouch as illustrated in Figure 6.10, which is quite handy for protection of valued objects. In recent years, PE foam penetrated into highway construction as concrete joints, into aerosteps for exercise, and shallow thermoforming for protection pads and utility boxes. Its cost and performance continue to find in-roads in different markets.

FIGURE 6.9
Packaging for a laptop computer using PE foam slabs. (Courtesy of Sealed Air Corp.)

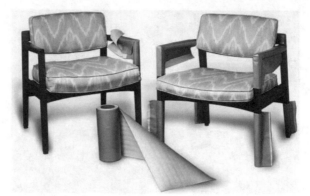

FIGURE 6.10
Object protection with foam coated with cohesive. (Courtesy of Sealed Air Corp.)

The x-linking enhances the inter-polymer strength, which allows for thin cell walls and fine-cell distribution. As a result, x-linked PE foams are soft, resilient, and have advantageous properties such as quick recovery and repeated usage. Thereby, they become suitable for applications that regular PE foams cannot entertain, such as shoe soles, helmet liners, pipe insulation, and so on. It also brings forth enough thermal strength for thermoforming with a shallow-to-medium draw ratio. In thermoforming, the temperature is much higher than the melting point of the polymer, but its high viscous nature brings forth enough thermal strength for a wide forming window. As a result, X-PE has been widely used in a variety of thermoforming applications such as knee pads, utility boxes, clam shell boxes, etc. In recent years, X-PE can be foamed in under 1 mm at 10 times expansion. The fine celled and soft nature, along with being somewhat cushioning, made it fine in body contact of medical treatment. The thin grade may bring extra mileage out of the X-linked structure. Therefore, various X-PE foam products in medical and sports applicationshave become available in the market.

X-linked slab, known as bun block, distinguishes itself in dimension, strength, and very fine cell structure. It is suitable for medium to high duty support. It can also be sliced into thin skin, or fabricated to certain geometry for insulation, support, and absorption usages in the areas of construction, automotive, and recreation.

Cross-linked PE foams can be made by either oven heat or the molded bead process. The latter involves irradiation of the pellets before the gas dissolution and foaming in the mold. This process is of batch nature consisting of x-linking, soaking, and foaming/fusion in a closed container to make the molded products. It can be extended to using linear polypropylene resins. A random copolymer is preferred. When fully expanded, the individual foam bead has a density as low as 12 kg/m^3, whereas after being molded into the final shape, the density increases because of the pressure on the beads to squeeze out the inter-bead space. A density increase up to 25% is very common, so that the final foamed product can have densities to about 16 kg/m^3. Because of the outstanding expandability, x-linked polyolefin foams have gained attention in different markets since the mid 1980s. The molded PP bead foam in small vehicle bumpers showed success in Japan in the 1990s.

In the late 1960s, conventional linear polypropylene grades have been converted into foams in the form of thin sheets using extrusion and the solvent flash process developed via a phase diagram approach [28]. The resulting foam density is very low, i.e., about 12 kg/m^3. It is thus excellent for packaging applications.

A new approach in making foamed PP began in the early 1990s. Medium-to-low density foams have been developed for thermoforming, packaging, and construction applications. However, since large quantities of blowing agent have to be introduced, blowing agent recovery became a necessary practice among processors to remain competitive. On the other hand, since conventional linear PP grades have a poor melt stability and are not capable of sustaining vigorous foaming, which is common in the low-density foaming practice, the advent of the polymer branching technology by irradiation or by reactive compounding (which increases the melt strength of the polymer and lowers its melting point), rendered the extrusion of low-density PP foams possible [172–179]. Since tertiary free particles are not stable, it is easy to cause chain scission than chain x-linking for PP. However, the temperature greatly affects the scission and x-linking ratio. Under 80°C, the x-linking rate outraces the scission rate, and a branched structure can thus be established.

PP has a melting point around 160°C. It becomes a natural extension of PE foam: melting from 100 to 130°C in temperature stretch. Another advantage is in the microwave foam tray. As mentioned before, PS foam is not suitable in this application. Although PP's modulus is not as high as PS, medium density foam and adding reinforcement layer (e.g., oriented PP film) made PP more and more popular in the market. On top of that, it has a very advantageous status in recycling and the environment. A medium density PP foam tray is portrayed in Figure 6.11. PP foam strands were tried to fuse into a slab via the extrusion process [179]. Its rigidity also opens doors in the automotive segment. More applications from PP are expected from its lamination with other structures to form a performance laminate or sandwiched composite.

Irradiated X-PP foam gained automotive market shares for its temperature resistance in recent years. Fine cell, thermoformed sheet found its way into liners, visors, and shields. With PVC skin, it can generate more usages in the interior of the vehicle. Its laminate with X-PE provides rigidity and soft feel for sports or body contact applications.

It should be pointed out that polyolefin blends have been found useful in modifying foam properties for a variety of applications. Blending linear or branched PP with low density PE can improve its mechanical properties and its thermal deflection. Importantly, the use of PP block copolymers with ethylene content can improve the compatibility between PP foams and PE surface. In particular, it is a common practice to enhance PE film and foam

The PP thermoforming process (schematics)

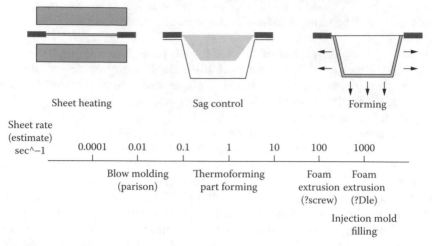

| Sheet heating | Sag control | Forming |

Sheet rate
(estimate)
sec^−1   0.0001   0.01   0.1   1   10   100   1000

Blow molding	Thermoforming	Foam Foam
(parison)	part forming	extrusion extrusion
		(?screw) (?Dle)

Injection mold
filling

FIGURE 6.11
PP foam tray thermoforming process.

TABLE 6.3

Polyolefin Foam: Methods and Markets

Name	Methods	Fabrication	Markets and Applications
LDPE foams	Extrusion	Cut, Lamination	Packaging, Floatation, Recreation
X-linked PE foams	Blending/heating	Thermoforming	Medical, Sports, Automotive
PP foams	Solvent/flash	Lamination	Packaging
PP foams	Extrusion	Thermoforming	Packaging, Recreation, Construction
c-PP foams	Bead Mold	None	Packaging, Insulation, Sports, Automotive
PE blend foams	Extrusion	Lamination	Packaging
Micrcellular HDPE foam	Extrusion	Blow Molding	Medical Bottles

lamination by using copolymer PP grades. Table 6.3 tabulates various foam technologies with their typical applications.

6.4 Other Foams

There are a number of other resins whose foams are developed in industrial applications. These include PET foams, PVC foams, epoxy foams, silicone foams, polyisocyanurate foams, Urea-formaldehyde foams, Fluoropolymer foams, etc. For the details, the readers can refer to other references such as reference #11. In this chapter, only PET and PVC foams are briefly mentioned.

Polyethyleneterephthalate (PET) resin can be made with a wide variety of material strengths for various applications. Normally, as the intrinsic viscosity (IV) increases, it requires a longer reaction time. For foaming operations, an adequate material strength necessary to hold the blowing agent molecules within the polymeric matrix is crucial for enabling the formation of low-density cellular structures. Conventional reaction processes do not seem to be economical enough in making foam grade PET. Branching, however, was recommended to reach the high IV for foaming. It can be established by introducing branching reactants to build up low crystallinity and highly branched units for melt strength. PET foams become noticeable in the tray market [180]. The inherent barrier property makes the foam intriguing in the packaging and food market.

PET is known for its great barrier property, good mechanical properties, and dimensional stability. Its thermoforming stability and food contact friendliness were recognized early. Since the 1970s, Crystallized PET (CPET) and amorphous PET (APET) became saturated in the worldwide food packaging market. Foamed PET added insulation to the packaging. In the early 1990s, foamed PET was first introduced to deep draw bakery trays Medium to high density PET foam continued making in-roads into food packaging. Microwave usage is a good example of foamed PET packaging. Due to its stiffness and temperature resistance, the foamed slab out of the branched PET shows potential in the areas of construction, transportation and automotive [181].

Polyvinylchloride (PVC) has been converted into a foam since the 1960s. It can be foamed through either conventional free expansion or constrained expansion known as the Celuka process [182,183]. It can also be foamed with or without plasticizers for achieving rigid or flexible characteristics. PVC's strength, stability, and insulation properties lend it to favorable applications, ranging from furniture and automotive, to construction and plumbing. It is easy to print and paint, and became a favorite in school projects, modeling materials, and signs. PVC foams were used in the core covered by thermoset epoxy or fiber glass into a sandwich structure. The closed cell PVC foam has a very low moisture absorption. Rigid PVC foam sandwich became the benchmark in naval architecture as illustrated in Figure 6.12. For PVC foam itself, its polar nature makes it very friendly in bonding with common adhesives or coatings. In decoration and interior/exterior design, PVC foams are very common. In the future, as long as its environmental status is properly set, it can easily penetrate into more and deeper applications.

FIGURE 6.12
Rigid PVC foam. Grid scored as core material for coating into a curved surface. (L.J. Gibson and M.F. Ashby, *Cellular Solids: Structure and Properties*, Pergamon Press, Oxford, 1988. With permission.)

7

Wood Composite Foams

7.1 Introduction

The main purposes of this chapter are to describe the applications of wood-fiber/plastic composites (WPCs), delineate the processing strategies and challenges associated with WPC foaming, and explain the state-of-the-art technologies used to produce WPC foams in various processes.

7.1.1 Outline

The purpose of Section 1 is to provide readers with general background information on WPC foams. This section will thereby introduce the concept of polymeric composites, the types of fillers used to make composites, the microstructure of wood fibers, and the polymers used in WPCs. The motivation for WPC foaming and the applications of WPC foams will also be discussed.

Section 2 explores the challenges to and processing strategies for the development of WPC foams. Sections 3–7 give accounts of the different foam manufacturing processes. Thus, Section 3 focuses on WPC foams in batch processing, while Sections 4 and 5 center on WPC foams in extrusion, Section 6 on WPC foams in injection molding, and Section 7 on WPC foams produced with an innovative stretching technology. Section 8 investigates the effects of nano-clay on WPC foams and their flame-retarding properties, and Section 9 is a chapter summary.

7.1.2 Polymeric Composites

A composite can be defined as a combination of one or more materials differing in form or composition on a macro scale. The constituents retain their identities; that is, they do not dissolve or merge completely into one another, but rather act in unison so that the desirable properties of each constituent are incorporated to create one product. This amalgamation thereby gives rise to a better material, one that offers an advantageous range of properties at a typically (but not always) lower cost. Normally, the components can be physically identified and exhibit an interface with one another.

Most polymer composites have two basic constituent materials: a binder or polymer matrix, and a reinforcement or filler. The matrix holds the reinforcements or fillers in an orderly pattern. The reinforcement is usually much stronger and stiffer than the matrix, and is generally comprised of either solid particulate or fibrous material. Property modifications can lead to reduced shrinkage, better directional strength, improved stiffness and scuff resistance, and enhanced electrical properties and dimensional stability [184–187]. In recent years, composites made of either thermoplastic or thermoset polymers that contain natural fibers as reinforcing agents have received increased attention in light of their superior properties and a climate of heightened environmental awareness [188,189]. Within the next decade, composites are expected to constitute the most important segment of the plastics industry.

7.1.3 Fillers in Polymeric Composites

Typically, two kinds of fillers are employed to make polymeric composites: mineral fillers and organic fillers.

7.1.3.1 Mineral Fillers

Mineral fillers have been widely used by the plastics industry for many decades. Currently, the most important mineral fillers are calcium carbonate, talc, silica, mica, clay, alumina trihydrate, and glass fibers [190–192]. Calcium carbonate ($CaCO_3$) has been quite popular for producing plastic composites because of its cost advantages and other appealing properties, including its low plasticizer absorption rate and strong resistance to thermal degradation during processing. Clays are useful in some specific applications, such as in the case of electric insulator products. Usually fillers are chosen with the aim of also decreasing material cost [193]. Another class of mineral fillers is known as "functional fillers," which are mainly used for improving the performance of the composite rather than for reducing its cost. They include glass fibers, glass spheres, carbon fibers, and fine-particle calcium carbonates, to name a few [194].

7.1.3.2 Organic Fillers

Although mineral fillers have many advantages (e.g., good heat, water, chemical, and electrical resistance), they impede the optimal functioning of composites in two important ways: they are characterized by a relatively high density and also lead to increased equipment abrasion. Organic fillers, such as WF, jute, flax, hemp, sisal, kenaf, pulp-fibers, bamboo fibers, rice-hulls, coconut husks, cotton husks, and sugar-cane husks, however, have a low specific gravity (about 0.9–1.3), and when used to extend resins, they offer significant cost savings [186, 195–197]. As compared to glass fibers, natural fibers have low tensile strength and comparable modulus, but their specific modulus (modulus/specific gravity) is often greater than that of glass fibers [195,197]. Wood-fiber predominates as the most extensively used fiber among all the cellulosic-fiber filler options.

A fiber is a long, fine filament of matter with a diameter generally in the order of 10 μm and an aspect ratio of length to diameter usually in the range of 100 to 1000. Fibers not only demonstrate strength, stiffness at the level of tension, and flexibility when they come to bending, but also often represent a combination of all of these qualities to yield a lighter product [198].

Wood-fiber/plastic composites (WPCs) have been commercialized and are replacing wood in a variety of building products, as well as in many automotive, infrastructure, and other consumer/industrial applications, due to their greater durability, improved dimensional stability, better resistance to moisture and biological degradation, low maintenance requirements, lower cost (compared to composites with carbon fiber, glass fiber, etc.), and better recyclability. However, the scope of their use in industrial applications has been somewhat limited because of their low impact strength and high density as compared to natural wood. Applying foaming technology to WPCs can decrease their density, and improve their mechanical properties (such as impact strength, toughness, and tensile strength), and improve their nail-ability and screw-ability all at a reduced material cost. To understand the crucial implications of foaming technology on WPCs, we need to learn first the microstructure and chemical constituents of wood and its fibers.

7.1.4   Microstructure of Wood Fibers

Wood itself is a naturally occurring, fibrous, and cellular composite, and is primarily composed of hollow, elongated cells that are arranged parallel to each other along the trunk of a tree. The fibrous cells, known as tracheids, have relatively thick walls and provide the strength necessary to hold the crown of the tree high up in the light. Wood, as constituted by these fibers, is composed of four basic constituents: cellulose, hemicellulose, lignin, and extractives [199,200]. Table 7.1 indicates the percentages, polymeric nature, and role of wood's chemical constituents [200].

Cellulose is a linear polymer composed of anhydro-D-glucopyranose units that form straight and stiff molecules, which exhibit a very high tensile strength [200]. Its affinity for water is the result of the large number of OH groups present on its chain. Hemi-cellulose is also a polymer made of sugar units, such as glucose, galactose, and mannose, for example, that similarly demonstrate a strong affinity for water. Lignin is a very complex, cross-linked, three-dimensional polymer formed from phenolic units. It provides

TABLE 7.1

Chemical Constituents of Wood

	% Composition	Polymeric Nature	Degree of Polymerization	Molecular Building Block	Role
Cellulose	45–50	Linear molecule (crystalline)	5,000–10,000	Glucose	Framework
Hemi-cellulose	20–25	Branched molecule (amorphous)	150–200	Primarily non-glucose sugars	Matrix
Lignin	20–30	Three-dimensional molecule	100–1,000	Phenolpropane	Matrix
Extractives	0–10	Polymeric	–	Polyphenols	Encrusting

rigidity to cell walls and hence to the overall wood structure, and is hydrophobic in nature.

Extractives comprise a category that includes all other chemical types, such as terpenes and their related compounds, fatty acids, aromatic compounds, and volatile oils. They are characterized by the fact that they can be extracted from wood when it is subjected to solvents. Extractives have no affinity for water and do not significantly contribute to the structural properties of wood. As mentioned above, water molecules attach themselves to the cellulose and hemi-cellulose molecules due to the presence of OH groups. But these water molecules are normally dislodged during the drying process, and when wood is said to have 0% moisture content (MC), it has a negligible or zero amount of water molecules attached to the OH groups on the cellulose [200]. Upon heating, wood releases moisture and other gaseous volatiles. The inherent moisture found in large quantities of wood-fiber may lead to the deterioration of the cell structure of foamed WPCs by negatively affecting cell size, cell distribution uniformity, and surface quality. Since these anomalies can therefore weaken the mechanical properties of foamed wood-fiber composites, the removal of moisture from the wood-fiber itself is a critical issue.

7.1.5　Materials Used for Wood-Fiber/Plastic Composites

Both thermosets and thermoplastics are used to make WPCs. Thermosets are plastics that cannot be melted by reheating once cured. These include resins such as epoxies and phenolics. The resins used in thermoplastic-wood composites, either recycled or virgin, are polyethylene (PE), polyvinyl chloride (PVC), polypropylene (PP), polystyrene (PS), and acrylonitrile-butadiene-styrene (ABS). Among these, PE has the major market share [201], followed by PVC, and then other resins like PP and PS. Because wood undergoes degradation quite easily, only thermoplastics that have melting temperatures below 200°C are commonly used in WPCs. Apart from this, the plastics are selected based on their properties, their cost, the manufacturer's familiarity with them, and the requirements of a given WPC application.

The wood used in WPCs is either in particulate form or very short fibers, otherwise known as "wood-flour." Long wood fibers that have an aspect ratio higher than 10:1 (or some other arbitrary ratio) are in fact those that are commonly referred to as "wood-fiber." However, no standardized, strict definitions exist for either term. This chapter invokes the term "wood-fiber" (hereon abbreviated as "WF") instead of "wood-flour," although both terms are used in the field of WPCs. WPCs typically contain up to 50% wood, although some WPCs contain as much as 70% (the latter often demonstrate increased processing difficulties, however). The common species used include pine, maple, and oak.

Some additives, such as coupling agents, light stabilizers, pigments, lubricants, fungicides, nucleating agents, blowing agents, and flame retardants, are typically added to either improve processing or enhance the performance of WPCs.

7.1.6 Advantages of Wood-Fiber/Plastic Composite Foams

Over the last two decades, interest in WPCs has been driven by a desire to decrease material costs and meet environmental regulations, as well as by advances in processing technology. However, the shortcomings of WPCs as an artificial wood—their heavy weight, low ductility, low impact strength, high flammability, and poor nail-ability and screw-ability—have limited their utility in many applications. Nevertheless, developing a fine-celled composite structure and incorporating flame-retarding additives into these composites can effectively compensate for these drawbacks. Furthermore, the foaming of WPCs results in lower material costs, better surface definition, and sharper contours and corners than when they are not foamed [202]. During production, the foamed composites run at lower temperatures and at faster speeds than their unfoamed counterparts due to the plasticizing effects of gas; consequently, the production cost is also reduced [202]. If the cell morphology of the foamed WPCs consists of a large number of uniformly distributed small cells, then the specific mechanical properties are also significantly improved [203,204].

WPC foams can be used in non-structural building products (decking, fencing, railings, window and door profiles, shingles, etc.), automotive products (interior panels, rear shelves, tire covers, etc.), and other numerous industrial and consumer products.

7.2   Challenges to and Processing Strategies for the Development of WPC Foams

7.2.1   Dispersion of Wood-Fiber and Fiber-Matrix Bonding

It is not easy to disperse WFs in the generally hydrophobic polymer matrix because of their polar structure. WFs tend to adhere to one another via inter-fiber hydrogen bonding due to the hydrophilic nature of their polar structure. Therefore, it is typically required that the fibers be treated chemically and/ or that external processing aids be added in order to facilitate the dispersion process. Coupling agents (also known as compatiblizing agents) are usually used to improve fiber-matrix bonding. These agents act to form chemical bridges between the resin and the fiber (or filler). Some of the most commonly employed coupling agents are organotrialkoxysilanes, titanates, zirconates, organic acid-chromium chloride coordination complexes, stearic acid, and maleated polymers. Coupling agents are especially useful in polyolefin-based WPCs because they overcome the incompatibility between the polar wood and the non-polar resin matrix. Instead, PVC-based WPCs do not necessarily need coupling agents to improve dispersion because both PVCs and wood have the same polar nature. However, it has been reported that silane-based coupling agents improve the mechanical properties of PVC/wood-fiber composites significantly [205,206]. In addition, processing equipment, such as K-mixers and twin-screw compounders [207], can also improve the dispersion process.

Apart from facilitating the dispersion of WF, coupling agents improve the mechanical properties of WPCs (i.e., its flexural strength, tensile strength, stiffness, and impact resistance) by enhancing the interfacial bonding between the WF and the plastic matrix. Coupling agents also improve dimensional stability, reduce creep, and decrease water absorption.

7.2.2   Processing Difficulties Due to Increased Viscosity

Generally, when the solid WF particles are added to the plastic resin, the viscosity of the molten mixture is increased. This effect becomes significant when the WF content is high. Figure 7.1 shows the non-Newtonian behaviors of WPCs with different WF contents (0%, 10%, 30%, 50%), as examined by a rotational rheometer. Ultimately, the viscosity increases significantly as the WF content is raised [208]. As a result, the processing pressure increases, causing certain processing difficulties. In turn, this necessitates that special adjustments be made to the equipment and die design. The greater the amount of WF added, the greater the number of associated problems, such as increased processing pressures and volatile content, as well as greater difficulty with the dispersal of WF in the polymer matrix. The processing pressure can be reduced by one of the following three alternatives: increasing

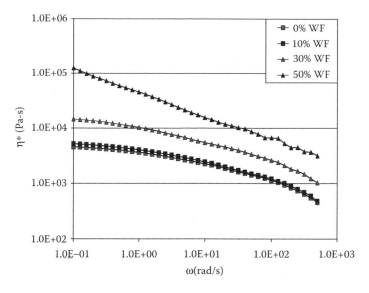

FIGURE 7.1
Complex viscosity vs. frequency for WPC with different wood fiber contents (0%, 10%, 30%, 50%).

the processing temperature, using a higher melt index (MI) material, or employing a lubricant package. However, at higher temperatures, a larger amount of volatile emissions is released from the WF, which subsequently deteriorates the cell structure of WPC foams [207]. Moreover, the temperature of the melt cannot be increased beyond 200°C in order for WF degradation to be kept to a minimum [207]. WPCs manufacturers tend to keep the WF content as high as possible to maintain reduced material costs. Thus, using a higher MI resin or using a lubricant package may be a viable strategy for increasing WF loading while keeping the processing temperature low. Additionally, if a resin exhibits a high melt strength, cell coalescence can be prevented; the final product will thereby retain most of the nucleated cells and consequently ensure a uniform cell morphology [209]. But the high MI resins have lower melt strengths, and are accordingly less able to accommodate a fine-celled structure. Hence, with the aid of a lubricant package, manufacturers should use resins of the lowest possible MI in order for processing to proceed at the desired WF loading.

7.2.3 Volatile Emissions Released from Wood-Fiber and Corresponding TGA Studies

As mentioned above, wood and WF are composed of four basic constituents: cellulose, hemicelluloses, lignin, and extractives [199]. Apart from these four constituents, wood also contains water. During high extrusion foam processing temperatures, WF releases moisture and other volatiles, leading to the

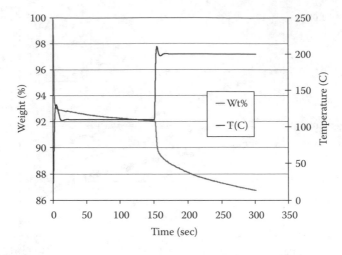

FIGURE 7.2
TGA thermogram for the devolatilization of WF.

deterioration of the cell structure of WPC foams by causing cell coalescence and cell collapse [209]. A thermogravimetric analysis (TGA) of the devolatilization behavior of the moisture and volatiles discharged by WF reveals that at 110°C (a typical drying temperature) and at 200°C (a typical processing temperature) (Figure 7.2), even after the WF is considered oven-dry, WF still emits approximately 3% volatiles when the temperature is raised from 110°C to 200°C [210,211]. Therefore, it can be deduced that in extrusion foam processing, whenever the processing temperature is raised to a higher level in any given zone of the barrel, additional moisture and volatiles will be generated and will in turn affect the foaming process significantly [207]. In order to yield a stable fine-celled morphology, the volatile contents of the WF particles need to be reduced to a bare minimum by adopting any of the standard drying techniques, such as online devolatilization [210,212], oven drying, hot air convective drying, drying in a K-mixer [199], and the like. During subsequent processing stages, the temperature should also be maintained below the drying temperature [207,210,213].

7.2.4   Critical Processing Temperatures in the Extrusion Foaming of WPC

In order to minimize the volatile emissions from WF during extrusion processing, the processing temperature should be kept to a minimum. However, lowering the temperature causes an increase in the "apparent" viscosity of the molten extrudate, and in so doing, poses greater processing challenges. The determination of an optimum processing temperature—one that does not compromise satisfactory processing conditions—is therefore crucial to ensuring the formation of acceptable cellular structures [207,213].

The highest processing temperature after the drying stage primarily governs the emissions from WF, which affect the foam morphology. Studies were conducted [213] to determine the critical processing temperature in a tandem extrusion system, as shown in Figure 7.3a [207], above which the excessive volatile emissions would prevent the formation of a uniformly distributed fine-celled structure. The drying method employed was online devolatilization, which occurred through a vent and without the use of a blowing agent—chemical or physical. The resultant foam structure was thus mainly generated by the WF emissions that were given off during processing. The typical cell morphologies of the composite foam samples (Figure 7.4) indicate that optimum composites were produced at a barrel temperature of 160°C; they exhibited small-sized bubbles and voids with uniform distribution. The cell morphology above this temperature was visibly irregular, and at lower temperatures, the foaming effect of volatiles from WF was insignificant .

7.2.5 Control of Residence Time in the Extrusion Foaming of WPC

As the TGA gives the time-dependant rate of weight loss at each condition, it can also be used to predict the amount of volatile emissions that may be released from the devolatilized WF upon further processing in the second extruder (Figure 7.3a) [207], which also impact the foaming process. It is assumed that the same processing temperature is used in the entire tandem system, and the volatile emissions liberated in the first extruder, in addition to the adsorbed moisture, are all purged at the devolatilizing vent or through the hopper inlet. It is also understood that most of the emissions released in the second extruder are trapped in the melt and contribute to the formation of the foam structure. The amount of volatile substances emitted in each extruder depends on how much time the composite melt spends in a given extruder.

Figure 7.5 illustrates the estimated amount of volatile emissions used in foaming based on the TGA results; these are essentially equivalent to the weight loss of WF (or volatiles released) in the second extruder. Comparing Figures 7.5 (a) through (d), it can be seen that as the residence time in the first extruder is increased, the volatile emissions in the second extruder are decreased. Moreover, the lower the residence time in the second extruder, the fewer the emissions from WF. This indicates that if WF is exposed to a high temperature for a long time in the first extruder, the amount of volatile substances generated from WF in the second extruder will be very small, even at higher temperatures (i.e., above 175°C). A knowledge of this strategy may possibly be used to produce fine-celled foam structures in WF composites with a high melting-temperature polymeric material. However, this would mean more degradation of WF in the first extruder. Therefore, it would be desirable to process the foamed WF composite materials at as low a temperature as possible, preferably below 160°C, as discussed earlier.

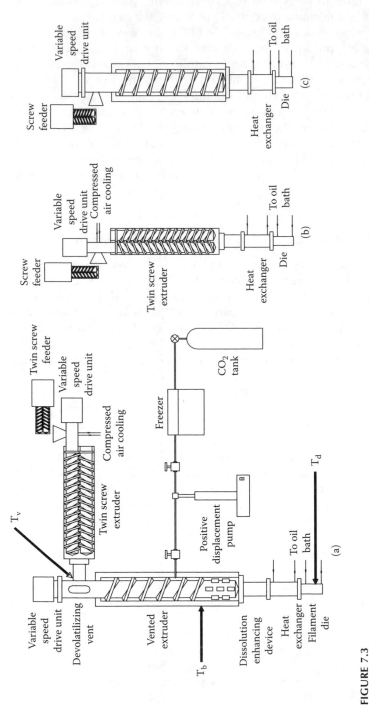

FIGURE 7.3
Schematic of WPC Processing Systems: (a) Tandem Extrusion System; (b) Twin-screw Compounding System; (c) Single-screw Extrusion System.

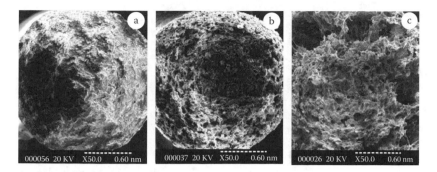

FIGURE 7.4
Effects of barrel temperature on the cell morphology of WF/HDPE foams. $T_v = 160°C$, $T_d = 135°C$; (a) $T_b = 145°C$; (b) $T_b = 160°C$; (c) $T_b = 175°C$.

7.3 WPC Foams in Batch Processing

WPC foams were first investigated using a batch processing technique [203–206]. First, compression-molded WPC samples were saturated with a physical blowing agent (CO_2 or N_2) in a high-pressure chamber. The saturated samples were then foamed by invoking thermodynamic instability via a pressure drop and a temperature increase, which is a similar process to the one invented at MIT for microcellular foaming (Figure 7.6). In batch processing, cell nucleation is governed mainly by the saturation pressure (or pressure drop), and cell growth is dictated by the heating temperature and time. Hence, the number of nucleated cells and the foam density can be controlled independently from one another. It must be noted that in a batch process, the foaming temperature is generally set as low as possible to make it easy to control the cell growth stage. As a research method, batch processing is extremely useful for obtaining critical processing parameters, such as foaming time, foaming pressure and temperature, and gas content. However, it is not preferable for the practical production of WPC foams since it is not a continuous process, and because it takes a long time to saturate the WPC samples due to the low rate of gas diffusion into the polymer at room temperature. Nonetheless, successful attempts at the microcellular foaming of WF/PVC composites using batch processing have been reported, as have 10 fold expansions in the fabrication of WPCs [203–206]. It has also been found that the solubility and diffusivity of gas were affected by the surface modification of WF with silane through sorption experiments using a batch processing system [203–206].

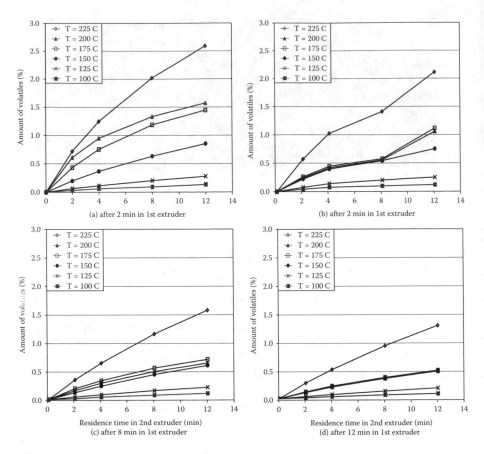

FIGURE 7.5

Estimated amount of volatiles used for foaming based on TGA data. The time in minutes indicates the residence time spent in the second extruder.

(a) After a residence time of   2 minutes in 1st extruder
(b) After a residence time of   4 minutes in 1st extruder
(c) After a residence time of   8 minutes in 1st extruder
(d) After a residence time of 12 minutes in 1st extruder.

7.4   Foaming of WPCs with Chemical Blowing Agents in Extrusion

7.4.1   Chemical Blowing Agents (CBAs)

CBAs, which are thermally unstable pure chemicals, are decomposed and released by blowing gas at a certain decomposition temperature. One advantage of CBAs is that they can be used either under high pressure or at atmospheric pressure. For example, rotational foam molding is conducted

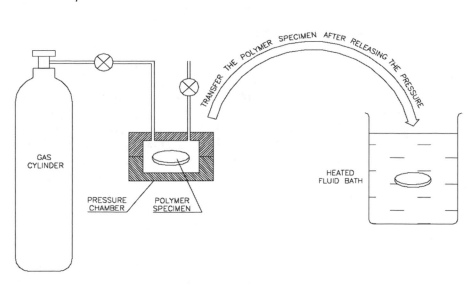

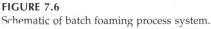

FIGURE 7.6
Schematic of batch foaming process system.

at atmospheric pressure. Compared with physical blowing agents (PBAs), CBAs are more expensive, but they do not require special storage equipment.

CBAs can be divided into two major categories: endothermic and exothermic. Exothermic CBAs release heat, while endothermic CBAs absorb heat during their decomposition. In general, endothermic CBAs generate CO_2 as the major gas yield. Commercially available exothermic CBAs primarily evolve N_2 gas, sometimes in combination with other gases. Compared to carbon dioxide, nitrogen is a more efficient expanding gas because of its slower rate of diffusion through polymers; it is thereby easier to control cell growth during the foaming process using nitrogen. It is worth noting that once an exothermic CBA begins to decompose, it continues spontaneously to do so until the CBA is entirely exhausted. This results in a faster decomposition across a narrower temperature range. In contrast, endothermic CBAs require additional heat to support their continuing decomposition. Consequently, they have a broader decomposition time and temperature range. CBAs employed in an extrusion foaming process [207] can be used either in their pure or masterbatch form (i.e., with or without polymer carriers). It has been reported that the polymer carriers in masterbatch CBAs can improve the compatibility between the CBAs and the polymer matrix, which may improve cell morphology [214]. Exothermic CBAs have been shown to produce a smaller cell size in wood composites than the endothermic CBA does in the extrusion foaming of wood/PVC composites [215,216]. However, it has also been reported that CBA types (endothermic vs. exothermic) and forms (pure or masterbatch) do not affect the void fractions achieved in both neat high density polyethylene (HDPE) and WF/HDPE composite foams [217].

When CBAs are being employed to produce WPC foams in extrusion, a few important factors need to be considered [207]. First, the decomposition temperature of a given CBA should be higher than the polymer melting temperature, but lower than the degradation temperature of WF, to ensure that the CBA will fully decompose to produce WPC foams without causing WF degradation. As mentioned before, the critical processing temperature should be kept as low as possible to suppress the generation of volatiles from WF. Therefore, one important criterion is to choose a CBA with a lower decomposition temperature. In addition to this, the rate of gas yield, gaseous composition, ease of dispersion, and cost should all be taken into account when choosing a CBA.

7.4.2 WPC Extrusion Systems

A single-screw extruder is the most common equipment used to produce WPC foams with CBAs. The low bulk density and stickiness of WF often cause feeding problems in an extrusion processing system; the WFs will congregate, creating a heap at the bottleneck of the hopper, while polymer pellets will pass smoothly into the extruder. Hence, the uniform feeding of a polymer pellet and WF mixture requires that the delivery rate be carefully controlled in order to prevent cramming in the feeding section of the processing system. The stable feeding of the polymer-WF blend is also integral to ensuring adequate bubble formation in the extrusion foaming of WPCs with CBAs [218]. Compounding the WF/plastic mixture is rather difficult because the hydrophobic plastic and hydrophilic WFs do not easily adhere to one another. The intensive mixing and uniform distribution of WFs in the plastic matrix is necessary for obtaining uniform properties in the desired WPC.

Typically, the hydrophilic WF and the hydrophobic polymer can be compounded using either a twin-screw compounder or a K-mixer, and then pelletized to yield WPC pellets, which are used for further extrusion foam processing [207]. The WPC pellets would prevent feeding issues, eliminate dusting problems that typically threaten the production environment, and evenly distribute WF throughout the plastic matrix. WPC pellets are commercially available on the market.

Certain patents have described the extrusion of foamed WPC profiles using the Celuka process [219–221]. In this procedure, the outer surface of the extrudate is cooled below the softening temperature of the plastic upon exiting from the die. The solidified skin prevents expansion in the outward direction; instead, the material swells inwardly, filling the hollow cavity that has been created by a solid torpedo inside the die.

The production of fine-celled WPC foams using CBAs in extrusion has been studied extensively [207,211,212,222,223]. Cell densities of the order of 10^7 were achieved; the largest cell sizes were less than 100 mm. Although microcellular structures have been generated around the WFs even with 50%

loading of WF, the volatile emissions from WF have prevented the formation of a uniformly distributed microcellular structure throughout the extrudate. The following sections will briefly present the extrusion process used to produce foamed WPCs using CBAs, as well as discuss the expansion behavior and cell morphology characteristic of these WPCs.

7.4.3 Two-Stage Extrusion Foaming of WPC with CBAs

Researchers have determined an optimal processing window for manufacturing WPC foams in extrusion with various CBAs whose decomposition temperatures are lower than 165°C [223]. WPCs are generally processed in two stages. In the initial stage, equal amounts (by weight) of HDPE and WF are mixed with a 3% coupling agent, and processed through a twin-screw compounder in which the temperature across all of the zones is raised to 175°C to effectively devolatilize the WF (Figure 7.3b) [207]. The extrudate is pelletized and then processed in a single-screw extruder (Figure 7.3c) [207] after an appropriate amount of CBA is mixed in.

7.4.4 Expansion Behavior of WPC Foams Blown with CBAs

It has been observed that the density of WPC foams, obtained without using any blowing agent, increases as the die temperature decreases in extrusion (Figure 7.7a) [223]. As the temperature of the die is lowered, the temperature

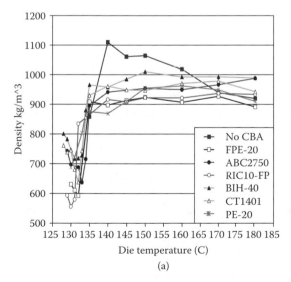

(a)

FIGURE 7.7
WPC produced with CBAs.

of the heat exchanger section is also decreased. This in effect reduces the amount of time that the WPC is exposed to higher temperatures, and thereby decreases the volatile emissions from the WF (a consequent increase in density is observed at lower die temperatures). As there are no, or very little, soluble blowing gases present, whatever volatiles are generated are lost easily without causing a density reduction. However, at the very low temperature of 135°C, some of these volatiles are successfully retained in the extrudate due to the freezing of the surface layer, resulting in a sudden drop in density.

Regardless of which CBA is employed, nearly all of the blowing agents display similar foaming behaviors. While it is expected that the density will start to decrease as the die temperature is lowered [211], there is, in fact, little change in the foam's density until a temperature of 135°C is reached. At lower temperatures, the diffusivity of the dissolved gas decreases so that more gas is retained in the extrudate as the temperature of the foam skin approaches the crystallization temperature. The dissolved gas continues to diffuse into the already nucleated bubbles, increasing their internal pressure and thus causing them to grow, which reduces the overall density of the WPC foam. However, as explained earlier, when the die temperature is lowered in order to decrease the foam's final density, the temperature of the heat exchanger, which is used as a diffusion-enhancing device, is also lowered. This subsequently reduces the amount of volatile emissions generated by the WF, as demonstrated in the case where a CBA was not used. At the same time, the diffusivity of the melt also drops so that more of the dissolved gases released from the CBA are prevented from escaping into the atmosphere and contributing to a decrease in density. Thus, there is a competition between a density decrease due to lowered diffusivity and a density increase due to a smaller evolution of gases from the WF. Apparently, the two competing effects approximately balance each other out, and the density remains nearly unchanged from 180 to 135°C, after which it drops suddenly.

The abrupt drop in density in Figure 7.7a suggests that the extrudate skin temperature quickly reaches the crystallization temperature upon exiting the die. At this point, the diffusion rate through the skin decreases sharply. The density reduction due to lower diffusivity at a lower temperature, as described earlier, should essentially be a gradual phenomenon following the rate of change of diffusivity with temperature. In contrast, the crystallized regions halt the diffusion of gas altogether. Due to the drastic reduction of gas loss though the foam skin at the crystallization temperature, nearly all of the trapped gas diffuses into the existing bubbles and increases the internal bubble pressure significantly, causing a sharp drop in the WPC's density.

It has been noted that WPC density increases again when the die temperature is brought below 130°C. As an extrudate surface temperature decreases further below the crystallization temperature, the amount of polymer being crystallized increases, which results in an increased stiffness in the polymer matrix. As a result, the extrudate resists further expansion and an increase in density ensues [224].

7.4.5 Cell Morphology of Extruded WPC Foams Blown with CBAs

Figure 7.7b shows a typical SEM picture. The cell density for all conditions was of the order of 10E7 cells/cm^3 and exhibited no variations on a logarithmic scale across different die temperatures.

7.5 Foaming of WPCs with Physical Blowing Agents (PBAs) in Extrusion

An innovative tandem extrusion system was developed for fine-celled foaming of WPCs using a PBA [207,225,226]. This system is capable of continuous on-line moisture removal and PBA injection (Figure 7.3a). It consists of two extruders: the first twin-screw extruder is used to compound the WF and the plastic, and the second extruder is used to foam the WPC. There is a vent at the connection of the two extruders that is used to remove moisture and volatiles online.

Compared to CBA-based processing, PBA-based processing (such as with environmentally friendly CO_2 and N_2) has no decomposition temperature requirements, reduces cost, and generally produces better cell morphology. Therefore, it is more logical to use PBA for WPC foaming, although it is technologically more challenging and necessitates proper capital investment (i.e., a gas injection system). The main feature of PBA processing is that PBAs (such as N_2 or CO_2) are injected into the extruder and uniformly dispersed into the plastic matrix under high pressure using high shear. The WF/plastic-gas solution is then homogeneously cooled down in a heat exchanger that has a static mixer in order to increase the melt strength, which is necessary to prevent cell coalescence. Finally, the WF/plastic-gas solution passes through the die, where foaming occurs [209]. WPCs with a fine-cell structure (i.e., cell sizes less than 100 μm) have been produced in extrusion foaming with N_2 [227]. It turns out that the PBA-based processing window is larger than the CBA-based one (Figure 7.8).

7.6 WPC Foams in Injection Molding

Injection molding is one of the most widely used plastic processing methods for mass producing complex parts which cannot be produced by extrusion. Research on foaming WPCs in injection molding is reported in [228–234]. The injection-molded parts in references [228–234] are made with wood-flour, PP and CBAs. The density of the injection molded WF/PP composites is reduced by about 24% using a CBA and the cell size varies from 10 to 50

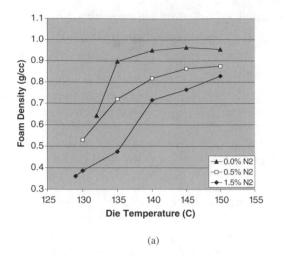

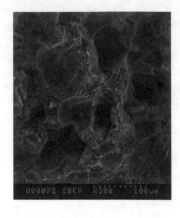

(a)   (b)

FIGURE 7.8
WPC produced with a PBA.

mm for the WF content ranging from 30% to 50% [229]. It has also been observed that the mechanical properties (specific tensile strength and specific flexural strength) of injection molded WF/PP composite foams improve up to 50% when the adhesion between the WF and the plastic matrix is aided by the use of a maleic anhydride PP-based coupling agent (5%). In addition, the injection molded composite foams without the coupling agent show remarkably higher water absorption than the composite foams with the coupling agent. This indicates that coupling agents would improve the dimensional stability of WPCs by reducing the water absorption. Bledzki et al. have also examined the effects of different CBAs and CBA contents, as well as the consequences of adding a coupling agent to the wood-fiber-PP foamed composites in injection molding and extrusion processing [233]. In their studies, injection molding yields better results than extrusion does, and the exothermic foaming agents perform better than the endothermic ones with respect to cell morphology and density reduction.

7.7   Innovative Stretching Technology and WPC Foams

7.7.1   Introduction

Recently, an innovative stretching technology was developed by Maine and Newson to orient polymer molecules and WFs in stretching, as well as induce the generation of voids at the WF and polymer matrix interface [235,236]. The advantages of an oriented WPC are that they demonstrate: (1) almost the same texture as wood; (2) the same density range as natural wood; (3)

excellent nailing and screwing abilities; and (4) better properties than un-oriented WPCs [235,236]. Molecular orientation is one of the beneficial results of the plastic deformation of polymeric materials. In most cases, it leads to an increase in the material's toughness and strength [237–239]. High-degree orientation of the polymer chain can be achieved by special processing techniques, such as ram extrusion (piston-cylinder), hydrostatic extrusion, and die drawing (solid-state extrusion) [240–242]. Since the temperature impacts chain mobility, the processing temperature in these procedures is usually maintained above the glass transition temperature but below the melting temperature. Below the glass transition temperature there is no chain mobility, and above the melting temperature, the chain mobility is high [240]. At certain low temperatures, the chains have limited mobility, and once stretched, they cannot curl up again. However, low temperature extrusion is very difficult to implement.

Coupling agents are usually used to disperse the WFs in the plastic matrix and improve the mechanical properties of WPCs by strengthening interfacial bonding. In addition to using coupling agents as a chemical means for making the surfaces of the WF and the plastic more compatible, the molecular orientation that occurs as a result of stretching can be employed as another way of improving the mechanical properties of WPCs. The stretching ability of WPCs is affected by their composition, the use of coupling agents, and the processing conditions. Typically, it becomes extremely difficult to use stretching technologies when the WF content rises above 40%. This is probably due to the increased brittleness of WPCs when the WF content is high. The following section will survey one batch processing case for stretched WF/PP composites with 30% WF [243,244] As a research method, batch processing is useful for determining important processing parameters.

7.7.2 Experimental Setup and Procedure for Stretched WPCs

A schematic of a WPC processing system is shown in Figure 7.9. The three-stages of WPC processing consist of melt-blending, extrusion and sizing, and stretching.

7.7.2.1 Melt-Blending of WPCs

Vacuum-oven-dried WF is dry-blended with PP pellets. The blended mixture is put into a high shear K-mixer (Werner and Pfleiderer Corp., Ramsey, NJ) and is melt-blended at 180°C. The melt-blended bundles are size-reduced to make granules using a granulator (C.W. Brabender). The vacuum-oven-dried granules are then fed into the counter-rotating intermeshing twin-screw extruder (C.W. Brabender: Model D6/2), which is driven by a 5 hp motor (Reliance Electric: Power Matched/RPM Rectified Power Motor). The mixture is plasticized and uniformly mixed by the intensive counter-rotation of

(1) Melt-blended PWC processing

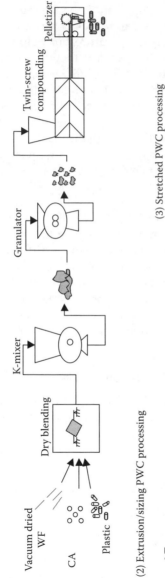

(2) Extrusion/sizing PWC processing

(3) Stretched PWC processing

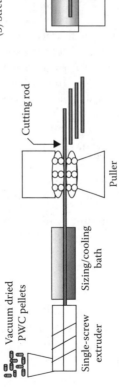

FIGURE 7.9
Schematic stretching of WPC processing.

the twin screws. The temperature is maintained below 175°C throughout the extruder in order to prevent the degradation of the WF.

The extrudate, which exits from the filament die, is cooled by blowing air over it. It is subsequently fed into a pelletizer, yielding WPC pellets of uniform size. Most of the vaporized water and other volatiles that are released during processing are purged into the atmosphere.

7.7.2.2 Extrusion and Sizing

Vacuum-dried WPC pellets are fed into the hopper of a single-screw extruder (C.W. Brabender: Model 3023-GR-8, L/D = 30). The temperature in this extruder is maintained at 150°C, 155°C, and 160°C in zones 1, 2, and 3, respectively. The WPC extrudate is passed through a nozzle die with a diameter of 7 mm. The die temperature is also maintained at 160°C. The extrudate, in its melt state, is cooled and solidified directly in a water-cooling bath while being pulled with a puller (Killion Extruder, Inc.: Model 2-12). By adjusting the puller speed, rods of different diameters are obtained. A puller speed of 4.5 m/min is adopted in order to make a rod with a diameter of 5 mm.

7.7.2.3 Stretched WPC Processing

Stretching experiments are carried out in a temperature-controlled oven (C. W. Brabender), which is connected to a stretching die and a puller. The stretching die is designed with a semi-angle of 10° and an exiting bore 3 mm in diameter.

In this batch processing of WPC, the effects of oven temperature and preheating time on the stretched products are investigated [243, 244]. The rods are preheated at various oven temperatures—110, 120, 130, 140, and 150°C—for different preheating times of 1, 2, 3, 4, and 5 hours, respectively, and then immediately stretched by a puller. A pulling speed of 35 cm/min is applied for all stretching experiments. The stretched samples are released from the stretching die and air-cooled immediately to prevent post-necking. It has been observed that when this precaution is not taken, "necking" occurs in the stretched samples [243, 244].

7.7.3 Effects of Preheating Time and Oven Temperature on Stretched WPCs

In the batch processing of WPC [243,244], the preheating time of the unstretched rods and the oven temperature should be considered to be able to result in a uniform temperature distribution throughout the stretched.

The resulting effects of the preheating time and the oven temperature on the density, as well as on the draw ratio of WF/PP composites with 30% WF, are shown in Figures 7.10 and 7.11, respectively. The density of the stretched samples dramatically decreases with the increased preheating time

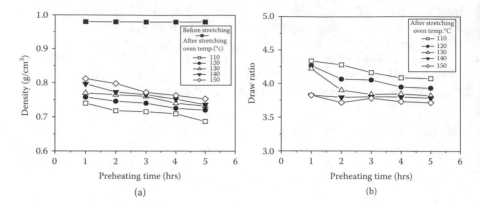

FIGURE 7.10
Effect of preheating time on (a) density and (b) draw ratio at various temperature for stretched WF/PP composites with 30 wt% WF.

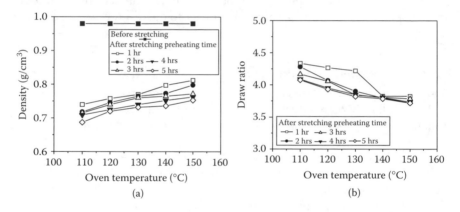

FIGURE 7.11
Effect of oven temperature on (a) density and (b) draw ratio over various preheating time for stretched WF/PP composites with 30 wt% WF.

at the various oven temperatures (Figure 7.10a). As the preheating time increases with each increase in oven temperature, the density of the stretched rods is reduced from 0.82 g/cm³ to 0.68 g/cm³. The unstretched rods, however, exhibit a density of 0.98 g/cm³. Thus, the stretched rod demonstrates roughly a 30% density reduction when compared with the unstretched rod. The drawing of composites lowers the density by generating voids because of poor interfacial adhesion between the WF and the PP matrix. Moreover, as the oven temperature is increased, the density of the stretched composites also increases (Figure 7.11a). Therefore, a lower temperature and a longer preheating time are favorable conditions for inducing density reduction.

The draw ratio decreased slightly as the preheating time and the oven temperature increased (Figures. 7.10b and 7.11b). This could have been

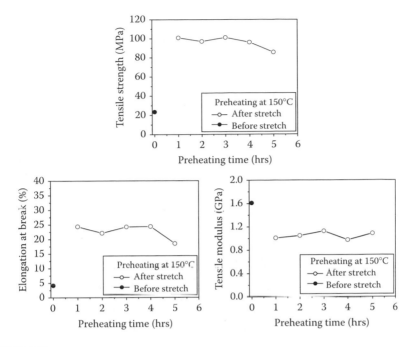

FIGURE 7.12
Tensile properties of stretched WF/PP composites with 30 wt% WF for various preheating time at 150°C.

caused by the necking in the die deformation zone, which is related to the strain rate field of the deforming polymer under drawing conditions. Draw ratio is a sensitive function of oven temperature but a weak function of preheating time. Therefore, a low temperature and a short preheating time are preferable for increasing the draw ratio.

7.7.4 Tensile Properties of Stretched WPCs

Figures 7.12 and 7.13 show how the preheating time and the oven temperature affect the tensile properties of stretched WPC when comparing the WF/PP composites with 30 wt% WF before and after stretching. As the preheating time of 150°C is increased, the tensile strength and the elongation at the break of the WF/PP composites decrease.

However, the tensile modulus (stiffness) does not change as the preheating time changes (Figure 7.12). The tensile strength and the elongation at the break of the stretched WF/PP composites with 30 wt% WF are 5 times as high as those of the unstretched ones, but the tensile modulus is lowered by 25% once stretching occurs. When the oven temperature is increased for those samples with a preheating time of 5 hours, the tensile strength decreases. The elongation at the break and the tensile modulus are slightly changed for this case (Figure 7.13).

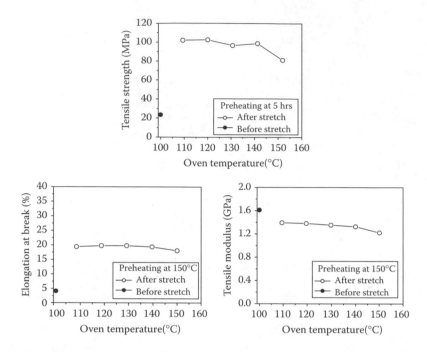

FIGURE 7.13
Tensile properties of stretched WF/PP composites with 30 wt% WF after preheating for 5 hrs at various oven temperatures.

FIGURE 7.14
Microscopes of stretched rod from the stretching die, showing (a) outside and (b) inside in the die deformation zone.

According to the above results, the tensile properties of the stretched WF/PP composites are not a sensitive function of the preheating time nor the oven temperature. The enhanced tensile properties are due to the polymer molecular orientation of the stretched WF/PP composites.

7.7.5   Morphology of Stretched WPCs

Figure 7.14 shows the stretched rod that has been removed from the stretching die. The rods demonstrate different colors prior to and after stretching. The original rod exhibit a brown color, but changes into an off-white color after passing through the die deformation zone. This results in a new product

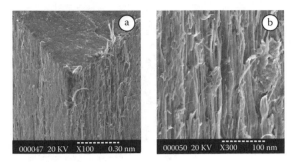

FIGURE 7.15
Morphology of stretched WF/PP composites with 30 wt% WF: (a) corner view, (b) enlargement of part of the side view of (a).

with a very similar wood-like appearance (Figure 7.14a). Figure 7.14b shows the morphology of the stretched rod that is cut in half along the machine direction (MD). Voids exist at the center of the stretched rod, which indicates that volatiles are generated and collected in the core area during the high temperature extrusion process. These voids are squeezed in the die deformation zone and are elongated along the direction of the machine.

Figure 7.15a illustrates the morphology of the two sides of a round rod section on the stretched WF/PP composites. Both the WF and PP matrix are aligned with the MD. As a result, the die drawing of WF/PP composites causes a unidirectional orientation of the polymer molecules and enhances the mechanical properties significantly along the MD. It seems that the orientation of the WF does not contribute to the mechanical properties because of the poor interfacial adhesion between the WF and the PP matrix.

Figure 7.15b is a magnified view of Figure 7.15a. It indicates clearly that WF particles are embedded in the elongated PP matrix channel. These fibers are not attached to the PP matrix because of a lack of adhesion at the interface between the PP matrix and the WF. They could, however, be easily removed from a cut surface by abrasion. The drawing of the composites obviously lowers the density by generating voids at the weak interface.

7.7.6 Effects of Coupling Agents on the Properties of Stretched WPCs

Figure 7.16 presents the effects of a coupling agent on the density of a stretched WPC. As expected from the function of a coupling agent, an increase in coupling agent content leads to an increase in the density of the stretched WPC. Because of the greater adhesion between the WF and the polymer matrix, the movement of WFs in the polymer matrix is diminished. As a result, the stretchability of the WPC is reduced and the formation of voids becomes difficult. Hence, the density of the stretched WPC increases.

Figure 7.17 shows the effects of the coupling agent on the tensile strength of WPCs. As the content of the coupling agent is increased, the tensile

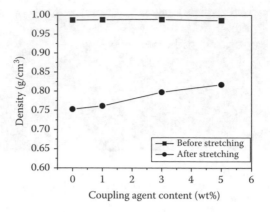

FIGURE 7.16
Effects of coupling agent on density reduction of stretched WPC with 30 wt% WF.

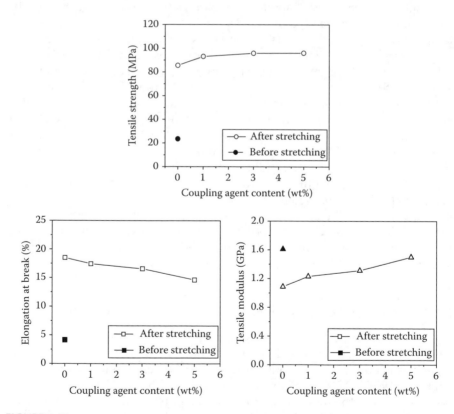

FIGURE 7.17
Effect of the coupling agent on the tensile properties of stretched WF/PP composites with 30wt% WF after preheating for 5 hrs at 150°C.

strength increases. However, once the coupling agent content is raised to 3 wt%, the tensile strength exhibits almost no increase. This suggests that 3 wt% may be the maximum and/or appropriate amount of a coupling agent needed for improving the tensile strength of a WPC. However, the elongation at the break decreases steadily because of enhanced interfacial bonding as the coupling agent content increases. The tensile modulus increases as the coupling agent content is augmented. It is believed that the fiber breakage mode is dominant when the interfacial bonding is enhanced with coupling agents [245,246].

In summary, the stretching of WPCs causes the unidirectional orientation of polymer molecules and enhances mechanical properties significantly along the stretching direction [243]. The use of a coupling agent can improve the mechanical properties of a WPC by strengthening the interfacial bonding between the polymer and the WF. However, the degree of void generation at the interface between the WF and the polymer matrix as a result of stretching decreases as the amount of coupling agent increases.

7.8 Effects of Nano-Clay on WPC Foams and Their Flame Retardancy

7.8.1 Introduction

The use of layered silicate nano-particles (i.e., clays) to reinforce polymers has drawn a great deal of attention in recent years [247–258]. Adding a small amount of clay can dramatically improve a number of polymeric properties, such as tensile modulus and strength, flexural modulus and strength, impact strength, thermal stability, flame retardancy, and barrier properties [247,248]. Thus, the introduction of clay into WPCs could be interesting from the perspective of improving their mechanical properties and flame retardancy, which are desirable effects, particularly with respect to construction applications.

7.8.2 Preparation of Polymer Nano-Composites with a Twin-Screw Compounder

In research conducted by Guo et al., mPE/clay nano-composites were prepared using a co-rotating intermeshing twin-screw extruder (W&P ZSK30, 40:1 L/D, D=30mm) [258]. This high shear screw is configured with neutral kneading elements and reversed conveying elements to produce a high shear and to increase residence time, thereby enhancing the mixing effect. The screw rotating speed was 50 rpm and the extrusion temperature was 150°C. The clay content was 5 wt% (Southern Clay Company, Cloisite 20A) and the coupling agent was 10 wt% (CA, Fusabond E-MB226D).

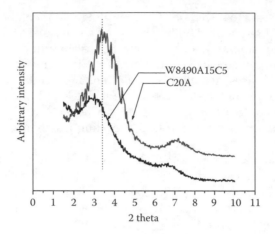

FIGURE 7.18
X-ray diffraction patterns for mPE/clay nanocomposites and clay.

The nano-composite structure that was obtained was characterized by wide-angle x-ray diffraction (WAXD). Figure 7.18 shows the XRD patterns of mPE/clay nano-composites used in this study. A weak peak is present at a lower angle for the nano-composites than for the pure clay. The pattern indicates that the nano-composites have an intercalated morphology in the mPE matrix. The nanocomposites were mixed with dried WF using a K-mixer, and were then pelletized in order to produce WF/mPE/nano-composite pellets for further extrusion foaming.

7.8.3   Effects of Nano-Clay on the Expansion Behavior of WPC Foams

Figure 7.19 demonstrates that when the nano-particles were introduced into the WPC for both the 30% WF and 50% WF cases, the density of the composites decreased regardless of the CBA effects. One contributing factor could have been that the nano-particles were well dispersed in the plastic matrix, thus increasing the melt stiffness and reducing the gas diffusivity. Because the gas diffusivity was reduced, the possibility of gas escape from the extrudate skin was also diminished. When 1% CBA was added in each case, the foam density was reduced. The mPE nano-composites, however, underwent a much larger density reduction than the mPE composites did. This is suggested by the bigger gap between the curve without CBA and the curve with CBA for the WF/mPE/nano-composite case.

7.8.4   Effects of Nano-Clay on the Cell Morphology of WPC Foams

The effects of nano-particles on cell morphology were investigated [258] (Figure 7.20). With the addition of 5 wt% nano-particles, the WPCs generally demonstrated a smaller cell size and more cells per unit volume. Although

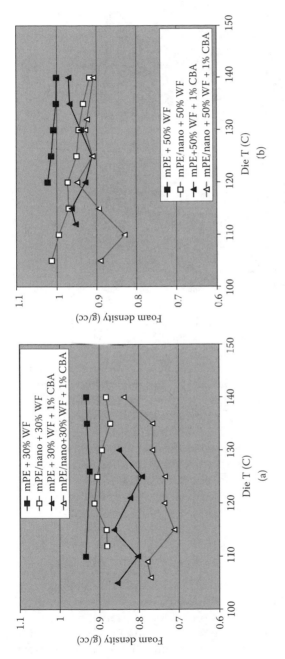

FIGURE 7.19
Foam density vs. die temperature: a) 30% wood fiber case; (b) 50% wood fiber case.

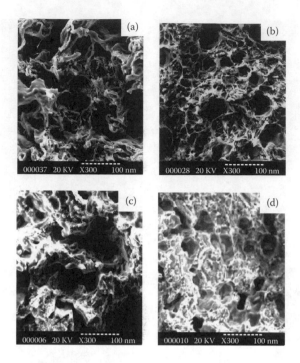

FIGURE 7.20
SEM pictures of mPE/WF composite foams mPE 30% WF 1% CBA at 125°C; b) mPE nano 30% WF 1% CBA at 125°C; mPE 50% WF 1% CBA at 125°C; d) mPE nano 50% WF 1% CBA at 125°C

the cell nucleation mechanisms for WPCs containing nano-particles have not been clarified yet, it is evident that, based on theories of heterogeneous nucleation, the addition of nano-particles increased the overall nucleation rate [250, 251]. In the heterogeneous nucleation scheme, the activation energy barrier to nucleation is sharply reduced in the presence of filler particles. This increases the nucleation rate and hence the number of bubbles produced. But further study is required to identify the exact mechanisms of cell nucleation (and cell deterioration, if any) for WPCs. The role of nano-clay in enhancing cell nucleation and/or suppressing cell deterioration also remains to be identified.

7.8.5   Effects of Nano-Clay on the Flame Retardancy of WPCs

A simple burning test was performed on WPCs with and without nano-clay, as shown in Figure 7.21. The WF/mPE/nano-composites showed enhanced fire retardancy. A piece of soft paper was placed on the table, and the rod foam samples were ignited and burned above it. For the WPC without clay, the burning sample dripped down quickly and ignited the paper. In contrast, the WF/mPE/nano-composites (5 wt% clay) formed a char during burning without dripping, thus retarding the burning and preventing the fire from

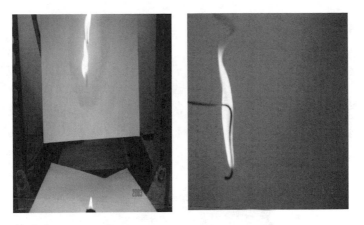

FIGURE 7.21
Burning test of WF/mPE/composites.

spreading. The flame retardancy tests (according to the ASTM D635) illustrate that the burning rate for the WF/mPE/nano-composites is lower than that for the WF/mPE/ composites (Figure 7.22).

7.8.6 Effects of the Degree of Exfoliation of Nano-Clay on the Flame Retardancy of WPCs

The effects of the degree of exfoliation of nano-composites on the flame retardancy of WPCs were investigated [259]. By applying different coupling agent contents, various HDPE nano-composites with small amounts (0.1–0.5 wt%) of nano-clay (i.e., conventional, intercalated, exfoliated) were successfully obtained using a melt compounding method in a twin-screw co-rotating

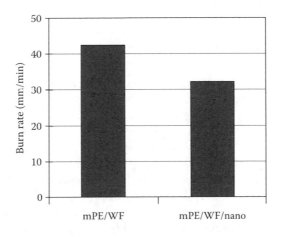

FIGURE 7.22
Burning rates of WPC and WPC with nano-clay.

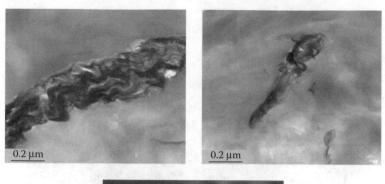

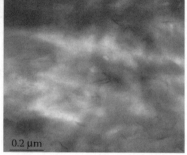

FIGURE 7.23
TEM for different HDPE nanocomposites at the clay content 1%. Conventional HDPE nano-composites; (b) intercalated HDPE nanocomposites; (c) exfoliated HDPE nanocomposites.

extruder. The morphology was probed using transmission electron micro-scopy (TEM). Figure 7.23 shows the TEM pictures of the conventional, inter-calated, and exfoliated HDPE nano-composites. Figure 7.23a indicates that the silicate clay particle in the conventional HDPE nano-composites did not exfoliate, while Figure 7.23b demonstrates that the silicate clay in the inter-calated HDPE nano-composites was partially exfoliated. Figure 7.23c shows that the silicate layers in the exfoliated HDPE nano-composites were fully exfoliated.

Figure 7.24 charts the burning rates as a function of clay content for wood-fiber/HDPE nano-composites made from conventional, intercalated, and exfoliated HDPE nano-composites. At a clay content of 0.1 wt%, the conven-tional and intercalated nano-composites had almost the same flammability as the composites without clay, while the burning rate of the exfoliated nano-composites decreased approximately 18% when compared with the wood-fiber/HDPE composites without clay. As the clay content increased, the burning rate decreased for all nano-composites, but the exfoliated nano-composites exhibited a sharper decrease. It is understood that achieving a higher degree of exfoliation of nano-clay is key to enhancing the flame retarding properties of nano-composites when a very small amount of clay is used. Gilman et al. reported that montmorillonite (MMT) clay must be

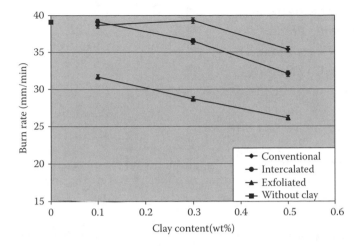

FIGURE 7.24
Burning rate vs. clay content for WF/HDPE nanocomposites made from conventional, intercalated, and exfoliated HDPE nanocomposites.

nano-dispersed to enhance the flame retarding properties of nano-composites [260]. In general, the flame retarding mechanism of nano-composites involves a high-performance carbonaceous-silicate char, which builds up on the surface during burning. This insulates the underlying material and slows the mass loss rate of decomposition products [260].

It also turns out that the same coupling agent used for helping with the dispersion of WFs in the polymer matrix and for improving the interfacial bonding between them, is also effective in the exfoliation of clay, and therefore no additional cost for including the nano-clay in the WPC is incurred.

7.9 Current Foaming Technology and Markets for WPC

Current foaming technologies can be found in many patents. Park et al. filed a patent [207] regarding foaming of WPC; Andersen Corporation manufactures foamed WPC products [261,262]; CertainTeed Corporation has one patent [263] regarding foamed polymer-fiber composites; Crane Plastics Company LLC (Columbus, OH) has one patent [264] regarding foam composite wood replacement material; Strandex Corporation [265] has developed its Strandex technology for WPC, and has licensed it to the following WPC companies:

Compos-A-Tron Manufacturing, Inc
Premium Composites, LLC
Louisiana-Pacific Corporation

PlyGem Industries, Inc.

Kroy Building Products

Universal Forest Products

Eidai Kako Co., Ltd.

Sekisui Chemical Co., Ltd.

STRANDEX EUROPE

Silvadec SA

There are still many other companies who are producing WPC foam products, and their production volume is not high but increasing.

7.10 Summary

A number of factors influence WPC foam processing when WF is used. The inclusion of large amounts of WF necessitates the use of higher MI resins or the use of a lubricant package. Coupling agents are used to enhance the dispersion and bonding of hydrophilic WF with hydrophobic resins. The emissions of volatiles from WF must be suppressed by drying the WF, placing restrictions on the highest allowable processing temperatures, and controlling the residence times at the highest temperatures. With the proper control of these factors, a uniform fine-celled WPC structure can be obtained. Active research is being carried out for producing WPC foams using a stretching process, which generates voids in WPCs and improves the mechanical properties of WPCs by orienting the molecular chains and the WF. The addition of nano-clay particles to WPCs increases foam expansion, improves cell morphology, and enhances mechanical properties, as well as reduces the flammability of WPCs.

8

Biodegradable Foams

8.1 Introduction

Interest in biodegradable foams is growing for a number of reasons. Increasing prices on plastic resins made from oil and natural gas have placed more emphasis on alternative materials. Further, the environmental concern of waste reduction has created a unique opportunity for the development of renewable polymers and using them for making polymeric foams. It is well known that the landfill problem associated with synthetic plastic products and shrinking space available due to population explosion has placed a greater emphasis on developing new polymeric materials that are either biodegradable or recyclable.

Biodegradable polymers are a new generation of polymers made from various natural resources that are environmentally safe and friendly. These polymers have potential applications in the packaging and non-packaging industry. There are a lot of studies still needing to be done to understand the potential impacts on the environment and recycling issues.

Biodegradable foaming and conventional thermoplastic foaming are different in many ways. First, resins from biodegradable foams are based on natural resources while resins from thermoplastic foams are petroleum-based. Biodegradable foams exhibit properties that are water soluble, degradable, and moisture sensitive whereas thermoplastic foams are non-water soluble, non-degradable, and do not exhibit much effect with moisture. The foaming process is similar, but the process for biodegradable foams must have low shear. For thermoplastic foams, the process is flexible and broad. Conventional foams with cell size ranging above 50–100 μm are possible to make from biodegradable plastics, but making a microcellular foam sheet (below 50 mm) with thickness above 10 mm and width above 1 meter is difficult with biodegradable polymers due to corrugation and equipment issues. More challenges are ahead for foam scientists. Water is polar in nature and mixes well with biodegradable resins such as starch. However, hydrocarbons mix well with some of the conventional thermoplastics such as polyolefins, but water has a compatibility problem. Therefore, the blowing

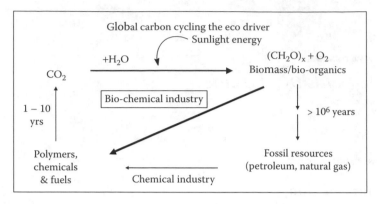

FIGURE 8.1
The global carbon cycle.

agent selection depends on the nature of the polymer, rheology, solubility, and diffusivity characteristics.

This chapter will outline the technology of making key biodegradable foams and their uses.

8.1.1   Importance of Biodegradable Plastics and Foams

Figure 8.1 illustrates the global carbon cycle for both the chemical industry and bio-chemical industry. By using annually renewable biomass feedstock rather than petrochemicals for the production of chemical and polymer needs, the rate of CO_2 fixation would increase and move towards equilibrium with the rate at which CO_2 is released. This would provide a better environment, better control and perhaps reduce the CO_2 emissions and help meet global CO_2 emissions standards specified by the Kyoto protocol, and provide a suitable alternative to the current polymer materials. Recent studies by Narayan [266,267] give more details on the biobased product drivers and polymeric materials. He also illustrates biobased and biodegradable materials and technologies as shown in Figures 8.1 to 8.3.

Direct extraction from annually renewable biomass feedstock is used to provide natural polymer materials such as starch, proteins, natural fibers, and cellulose. This is shown in Figure 8.2. On the other hand, Figure 8.3 shows that biomass feedstock can be converted to biomonomers by fermentation or hydrolysis and then, further converted by chemical synthesis to biodegradable polymers.

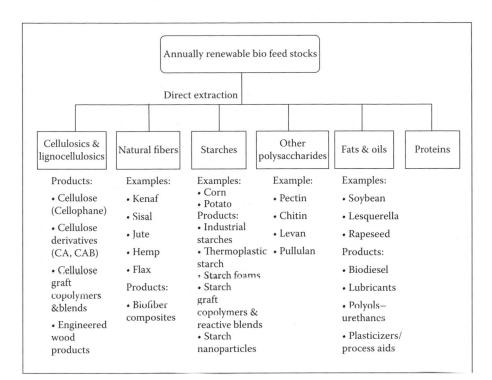

FIGURE 8.2
Direct Extraction from Bio Feed Stocks.

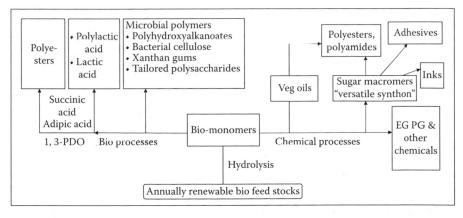

FIGURE 8.3
Conversion of Biomass Feedstock to Biopolymers.

8.2   Definition of Biodegradability

The term "biodegradable polymer" refers to the susceptibility of a polymer to decomposition by living things or by environmental factors. The American Society of Testing Materials defines biodegradable as capable of undergoing decomposition into carbon dioxide, water, methane, or biomass resulted from the enzymatic action of microorganisms, which can be measured by standardized tests, in a specified period of time reflecting available disposal condition. It is important to understand that this definition requires a time limit. Natural environmental factors that cause decomposition include bacteria, fungi, molds and yeast. The degradability is defined as the ability of materials to break down by bacterial, thermal or ultraviolet (UV) light action. Degradation of polymers by UV light is also called photo-biodegradation, in which the polymer degrades into low molecular weight material and later is converted to carbon dioxide and water by bacterial action.

8.3   Biodegradable Polymers

A wide range of biodegradable plastics is available on the market. Table 8.1 lists some of the important polymers.

Biodegradable polymer offers a wide range of products. It can be made into thin films for making shopping bags, composting bags, mulch film and landfill covers. It can be foamed to make packaging foam and biodegradable food service boxes and cups. While biodegradable foams offer several advantages it is important to avoid the risk of contamination by conventional plastics such as polystyrene and polyolefins which are normally reprocessed or recycled. The sorting methods and techniques are still in infancy and more research and resources are needed to detect and differentiate biodegradable polymers from others.

TABLE 8.1

Biodegradable Plastics & Applications

Biodegradable Plastics	Foams/Films	Application
1. Starch and its blends	Loose fill foam	Packaging
2. Biodegradable polyesters - Polylactic Acid (PLA) & others	Sheets	Thermoformable sheets
3. Polycaprolactone (PCL)	Foamable	Packaging
4. Polyvinyl Alcohol (PVOH)	Water soluble foam or film	Packaging
5. Ethylene vinyl Alcohol (EVOH)	Water soluble foam or film	Oxygen barrier, multi-layer packaging and foams

8.4 Advantages and Disadvantages

Biodegradable foams offer several advantages over conventional thermo-plastic foams. For example, the disposal of biodegradable foam to landfill may accelerate waste degradation decreasing landfill space usage. Foams made from polyvinyl alcohol are water soluble in nature and that improves the environmental friendliness in handling and disposing of the product. Although there are several advantages claimed, there are some disadvantages and environmental risks associated with biodegradable plastics and foams. For example, there is a chance of plastic additives, plasticizers, modifiers and colorants washing off from landfills and mixing with groundwater to contaminate surface or well water sources. There may be sensitive fish and other species affected due to partial degradation of biodegradable plastics including films and foams in marine environments. More research studies are needed to understand environmental problems associated with the use of biodegradable plastics. Particularly, it is essential that clear disposal plans, processing methodologies, testing, life-cycle analysis and educating the public are to be focused on to achieve success in using biodegradable foams for new market applications.

8.5 Degradation Mechanisms

It is important to understand the different terminologies used in referring to biodegradable plastics. Over many years of careful research, much of the confusion surrounding biodegradable polymers was eliminated. Many research groups have established benchmarks and ASTM standard to assess biodegradability and compostability. For example, Table 8.2 shows the major classification and degradation mechanisms explained by ASTM test method. A detailed list of biodegradable plastics developments and environmental impacts were reported in 2002 [268].

8.6 Development of Biodegradable Foams

Foamed polymers offer unique advantages such as reduced weight, better thermal insulation, low material usage, cushioning, sound absorption and packaging benefits. Typically biodegradable foams are made by the conventional extrusion technique. Table 8.3 shows key developments in this area of research.

TABLE 8.2

Classification of Biodegrable Polymers

Classification of Biodegradable Polymers	Mechanism	Reference
1. Biodegradable Polymer	Degradation results from the action of naturally-occurring microorganisms such as bacteria, fungi and algae	ASTM D6400
2. Compostable Polymer	Degradation results by composting to yield carbon dioxide, water, inorganic compounds and biomass and leaves no visible, distinguishable or toxic residue	ASTM D6400
3. Hydro-Biodegradable Polymer	Polymer breaks down by initial hydrolysis followed by further degradation	—
4. UV-Biodegradable Polymer	Polymer breaks down by UV light followed by further degradation	—
5. Bio-erodable polymer	Oxidize and degrade under weather conditions	—

TABLE 8.3

Historical Development of Biodegradable Foams

Year	Development	Authors	US Patent Number
1964	Extrusion of Starch to produce an expanded gelatinized product	Protzman, Wagoner	US 3,137,592
1975	Extrusion of hydrophobic porous starch product	Boonstra, Berkhout	US 3,891,624
1989	Biodegradable packaging starch foam	Lacourse, Altieri	US 4,863,655
1991	Biodegradable shaped products	Lacourse, Altieri	US 5,035,930
1992	Water soluble PVOH foams	Malwitz, Lee	US 5,089,535
1996	Improved water/humidity resistant starch foam	Altieri, Tessler	US 5,554,660
1996	Starch foams for absorbent articles	Griesbach	US 5,506,277
1998	Biodegradable foam-expanded material	Kakinoki, Sato	US 5,766,749
1998	Biodegradable foam moldings of thermoplastic starch	Schennink	WO 9851466
2001	Foamed starch sheet	Bastioli, Lombi, Salvati	WO 0160898
2002	Expanded biopolymer-based articles and process of producing	Van Tuil, Van Heemst	WO 0220238
2002	Biodegradable foamed product from starch	Fisk	US 6,406,649
2004	Hydroxycarboxylic acid-copolymer for disposable material (including foams)	Bigg, Sinclair, Lipinsky, Litchfield, Allen	US 6,740,731

8.7 Extrusion of Novel Water Soluble PVOH Foams

8.7.1 Polyvinyl Alcohol (PVOH) Resin Chemistry and Rheology

8.7.1.1 PVOH Chemistry

PVOH is generally a non-thermoplastic material, which degrades easily to a yellow powder upon heating before reaching its melting point. To make it thermoplastic, polyols can be added to reduce its melting point to improve processing. Another alternative is to use copolymerize PVOH to plasticize the material. PVOH is a polymer with pendant hydroxyl groups on alternating carbon atoms. These hydroxyl groups promote inter- and intra-molecular hydrogen bonding which affect physical properties of PVOH such as water solubility and mechanical properties such as tensile strength and elongation, the glass transition temperature which in turn affects extrusion processing characteristics. The break-up of the crystalline structure is a necessity to achieve processability in extrusion operations such as foaming and film blowing.

The melting point of PVOH is a function of its ester content, its molecular weight and plasticizer content. For example, if the ester content is high then it disrupts the crystallinity and hydrogen bonding more then the melting point becomes lower. Similarly the addition of plasticizers such as polyethylene glycol, glycerol and neopentyl glycol reduce the crystallinity, as a result, the polymer has thermoplastic behavior.

8.7.1.2 Manufacture of PVOH

PVOH cannot be made from PVOH monomers since they do not exist in the free state. So it is manufactured from polyvinyl acetate by hydrolysis. The hydrolysis is accomplished by reacting polyvinyl acetate with methanolic sodium hydroxide and the properties of the resulting PVOH depends on the structure of the precursor polyvinyl acetate. Depending on the degree of hydrolysis, a wide range of materials can be produced with different levels of water solubility and performance. In general, the properties of PVOH polymer depend upon the degree of hydrolysis, level of crystallinity, molecular weight and its distribution, crystalline structure and orientation, and ambient temperature and humidity. The following list shows the effect of molecular weight and hydrolysis level on the mechanical properties of PVOH.

Increase in Molecular Weight & Percentage of Hydrolysis:

- Higher viscosity
- Increased tensile strength
- Increased melt strength

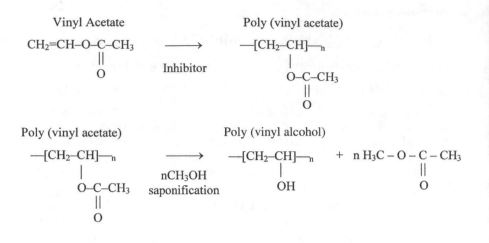

- Increased water and solvent resistance
- Better adhesion to hydrophilic surfaces

8.7.1.3   Vinex PVOH Thermoplastic Copolymer Resins

Air Products and Chemicals introduced various commercial resins called Vinex resins which are basically thermoplastic PVOH copolymers. They were produced by using an internal plasticizer. This is achieved by conducting the polymerization reaction in the presence of comonomers and grafting reactions. They are water soluble and can be processed or foamed by a conventional extrusion process. U.S. Patents 4,618,648 and 4,675,360 discuss more details. Vinex resin is a copolymer of PVOH with poly(alkyleneoxy) acrylate. The function of poly(alkyleneoxy) acrylate is to break up the crystallinity of PVOH to lower the glass transition temperature of the polymer, lubricate the polymer in the melt, and retain cold water solubility. The chemistry and technology of this development led to the introduction of Vinex resins, which are internal plasticized, thermoplastic and water soluble polymers. Several grades of Vinex grade resins were available. For example, the following typical properties were reported for Vinex 4004 and Vinex 4025 resins as shown in Table 8.4. There are other grades like Vinex 2034 (88% hydrolysis) and Vinex 1003 (99% hydrolysis) that are available for extrusion foaming. The above resins can be blended to make foams of interest.

A typical 1 to 1.5 mm thick film made out of Vinex 4025 resin broke up in 14 seconds and dissolved completely in 25 seconds in distilled water at 25°C under slight agitation.

8.7.1.4   PVOH Resin Rheology

Figure 8.4 shows PVOH resin rheology data for Vinex 1003 which is 99% hydrolyzed. This data shows the resin has good melt viscosity characteristics

TABLE 8.4

Properties of Vinex 4004 and Vinex 4025

Properties	Vinex 4004	Vinex 4025
Melt Index	4.5	14
Melting Range, °C	165–210	165–195
Glass Transition Temperature, °C	46	24
Melt viscosity, poise at 195°C	9408	2970
Tensile strength of film @ 23°C and 50% RH	4300 psi	3500
Elongation of film @ 23°C and 50% RH	50%	250%

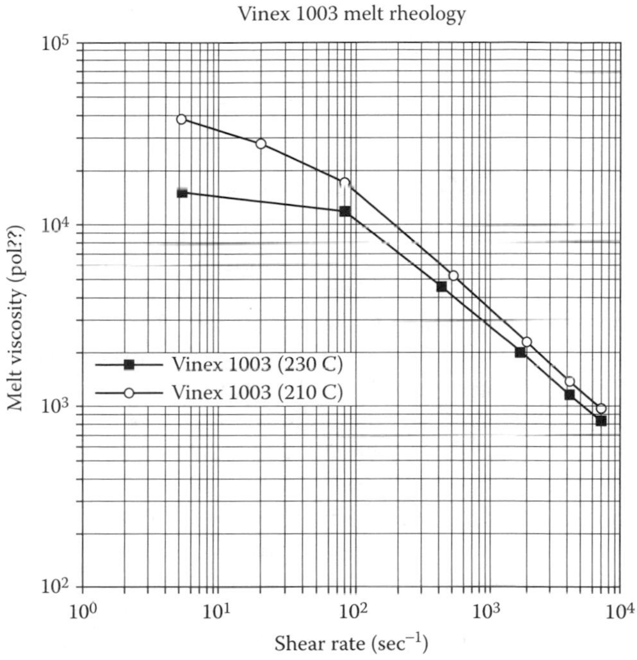

FIGURE 8.4
Melt Rheology of Vinex 1003.

suitable for foam processing. The shear viscosity is relatively flat up to 100 sec^{-1} shear rate and then shear thins significantly. This is ideal for foaming because it can run without adding much load in the extruder, but offer good melt strength during bubble growth outside the die where shear rate is almost zero.

Figure 8.5 shows the effect of blending Vinex 1003 in Vinex 2034 on the foam density. Vinex 2034 is another PVOH blend resin that has a melt index of 4 MI. Changing the Vinex blend ratio can change the resulting density of the foam.

Recently there is a great interest in industries to develop plastics from water soluble and biodegradable such as polyvinyl alcohol (PVOH). PVOH

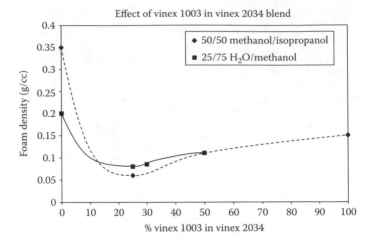

FIGURE 8.5
Foam Density vs. Vinex 1003 in Vinex 2034 Blend.

is prepared by hydrolyzing polyvinyl esters, e.g., polyvinyl acetate. The advantage of using PVOH is that the resulting solute is biodegradable. Ramesh and Malwitz investigated extrusion of PVOH foams [269].

With the phasing out of CFCs due to their high ozone depletion potential and future restrictions on hydro chlorofluorocarbons the exploration of alternative blowing agents for foam processing has been increased.

This technology deals with the development of a reliable continuous extrusion process to foam polyvinyl alcohol by using methanol and water. The processing polyvinyl alcohol has been severely limited by the fact that it decomposes before melting. Extrusion process requires a stable "melt" of the polymer during processing. Until recently, means for melt extruding materials processing the desirable characteristics of PVOH have not been available. The later part of the paper describes a heat-transfer model coupled with viscoelastic foam growth model with modified boundary conditioned in predicting the final foam density.

8.7.2   Experiments

Experiments were conducted using a co-rotating twin-screw extruder. The extruder used was highly instrumented and computer interfaced during operation. The blowing agent was injected at the mid-point of the barrel where the co-rotating screws have reverse flight which assures a tight melt seal under processing conditions. The screws are partially intermeshed, with their positive pumping action, can easily provide the necessary pressure to create a tight seal at any operating condition. A melt seal impedes the flow of blowing agent backward toward feed opening. Figure 8.6 shows a

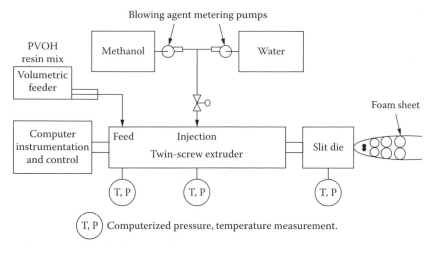

FIGURE 8.6
Schematic diagram of the Twin-Screw Extrusion Foaming Process.

TABLE 8.5

Extrusion Compositions and Conditions of Water Soluble Foams

	Formulation		
Ingredients	1	2	3
Solids			
Polyvinyl Alcohol Resin (Vinex 2034)	89.47	88.55	86.32
Nucleating Agent	1.30	1.30	1.30
Cross linking agent	0.3	0.30	1.0
Liquids (Blowing agent)	8.93	9.85	11.38
Methanol & water mixture (maximum)			
Foam Density (kg/m³)	53.60	44.85	35.24
Extruder			
Melt Section = 505K			
Injection Section = 455K			
Die Section = 408K			
Die Pressure = 1400 psi			

schematic diagram of the extrusion setup. Table 8.5 shows the extrusion conditions for various formulation runs.

8.7.3 Computer Model

When the foam sheet exits the die, heat conduction is the primary mechanism as pointed out earlier by Lee and Ramesh [270].

Although a very slow conduction of heat to ambient air is expected during foam expansion because of poor conduction of polymer and gas, the heat transfer effects are important especially when thin sheets are extruded. For thin sheet, it is reasonable to assume an average temperature flat profile as described in the literature [270]. The average temperature of the thin sheet is given as:

$$T_{avg} = T_s - \frac{8}{\pi^2}(T_s - T_o)\exp\left(-\frac{\alpha\pi^2}{l^2}\right)$$ (8.1)

where α is the thermal diffusivity, $\alpha = k/\rho c_p$. T_o and T_s are the initial temp of the sheet (which is equal to the die temperature) and surrounding room temperature respectively. Variables "t" and "l" denote foaming time and sheet thickness. To account for the high void percentage during foam growth, a foam packed thermal conductivity coefficient was employed as shown previously [270]. Arrhenius type viscosity expression was used to account for the viscosity changes as shown below.

$$\eta_o^* = \exp\left[\frac{E_v}{R_g}\left(\frac{1}{T} - \frac{1}{T_o}\right)\right] \cdot f(c)$$ (8.2)

where E_v/R_g is the activation energy of viscosity determined experimentally, which is 11.13 kcal/gm mole K for polyvinyl alcohol system.

In order to characterize the viscosity of the polymer with blowing agent mix, rheological experiments were conducted using Haake's on-line capillary rheometer as illustrated in Figure 8.7.

Since it is a closed system, it is ideal for the characterization of viscosity of blowing agent with polymer mix at various blowing concentrations. The ratio of the viscosity of the polymer with the blowing agent to the viscosity of the pure polymer, f(c), is also referred to as a viscosity reduction ratio in

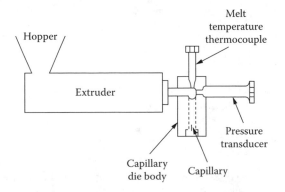

FIGURE 8.7
Schematic diagram of Haake's Rheocord 90 Extrusion Capillary Rheometer.

the literature [271]. The experimental values of f(c) were determined using the Haake on-line rheometer.

Besides viscosity, diffusivity and solubility are affected by the changes in temperature. Diffusivity, D, increases with an increase in temperature. On the other hand, solubility decreases with increase in temperature. They are expressed as follows [33].

$$\log D(T) = -5 + \left(1 - \frac{435}{T}\right) \cdot 10^{-3} \frac{E_D}{R_g} \tag{8.3}$$

where

$$\frac{E_D}{R_g} \cdot 10^{-3} = \left(\frac{\sigma_x}{\sigma_{N_2}}\right) \left[7.5 - 2.5 \cdot 10^{-4} \left(T_g - 298\right)^{3/2}\right] \tag{8.4}$$

$$D = D(T)(1 - x_c) \tag{8.5}$$

where x_c is the degree of crystallinity; σ_x, σ_{N_2} are collision diameter of the diffusing gas and nitrogen respectively; "T_g" is the gas transition temperature of the polymer. The individual diffusivity of methanol and water (D_{OH} and D_{water}) can be calculated from Equations 8.3 to 8.5 and then can be substituted into Equation 8.6 for calculating the effective diffusivity of the mixture. This is the best way to correlate the diffusion of mixed composition gases for foam expansion. Because a mixture of methanol and water is used as blowing agent, the individual calculated parameters should be related using the standard mixing rule according to Wilke [272].

$$D = D_{mixture} = \frac{1}{\left[\dfrac{y_{OH}}{D_{OH}} + \dfrac{y_{water}}{D_{water}}\right]} \tag{8.6}$$

where y_{OH} and y_{water} are mole fractions of alcohol and water respectively. The required physical constants are given in Table 7.5. Similarly, for solubility the equations are:

$$\log k(T) = \log k_o - \frac{435}{T} \cdot \left(\frac{\Delta H_s}{R_g} \cdot 10^{-3}\right) \tag{8.7}$$

where

$$\log k_o = -6.5 - 0.0057\left(\frac{\varepsilon}{\kappa}\right) - 0.013\pi \tag{8.8}$$

$$\frac{\Delta H_s}{R_g} = 450 - 13.25\left(\frac{\varepsilon}{\kappa}\right) \tag{8.9}$$

$$k_w = k\left(1 - x_c\right) \tag{8.10}$$

where k_w is the Henry's law constant, $\Delta H_s/R_g$ is the molar heat of sorption, and (ε/κ) is the Lennard-Jones temperature.

The concept of a cell model was developed for studying the growth of loosely spaced bubbles in Newtonian fluid by assigning a calculated volume of polymer based on cell nucleation [112].

Later it was modified to include the viscoelastic effects [273] and tested against the experimental data of foam growth in polystyrene [274]. Recently more modifications were made to include the transient heat transfer effects [270]. More details of the model and equations were published elsewhere [269].

It should be noted that previous studies in the literature [112, 274] assume that there is no gas loss from the bubble during the growth process. This condition is not exact when considering the sheet extrusion process where a large surface area of the foamed sheet is exposed to the atmosphere and gas loss occurs to the atmosphere from the foam sheet. A finite difference technique was used to solve the above set of equations to determine the radius of the growing bubble as a function of time. The bulk foam density can be calculated from the following equation:

$$\rho_{bulk\text{-}foam\text{-}density}\left(t\right) = \frac{mass \cdot of \cdot the \cdot unit \cdot cell}{vol. \cdot of \cdot bubble \cdot at \cdot time \cdot 't' + vol. \cdot of \cdot concentric \cdot envelope} \tag{8.11}$$

The diffusion and solubility constants and all other required parameters were obtained using molecular group contribution theory [275]. The physical constants used in the simulation are given in Table 8.6.

8.7.4   Results and Discussion

8.7.4.1   *Effect of Cross-Linking Agent*

It should be noted from Table 8.1 that the maximum possible concentration of blowing agent for producing non-collapsing foam for run #2 is 9.85% by weight. The product density was about 44.8 kg/m³ (2.8 pcf). When further

TABLE 8.6

Physical Properties for Polyvinyl Alcohol-Methanol-Water System [276]

No.	Physical Constants	Values	Units
1	Surface tension of polyvinyl alcohol (σ)	357×10^{-3}	N/m
2	Collision diameter of methanol (σ_{OH})	36.3	N/m
3	Collision diameter of water (σ_{water})	26.4	N/m
4	Collision diameter of nitrogen (σ_{N2})	38.0	N/m
5	(ε/κ) for methanol	482	K
6	(ε/κ) for water	809	K
7	Permachor value for PVOH (π)	157	--
8	Degree of Crystallinity	0.25*	--
9	Density of Polyvinyl alcohol	1250	kg/m^3
10	Density of methanol	790	kg/m^3

* from material supplier's data

blowing agent was added the foam started to collapse due to shear thinning effects. To achieve lower density, the concentration of cross-linking agent was increased from 0.3% to 1%. Due to cross-linking effect the overall viscosity of the melt increased significantly. This allowed the addition of more blowing agent to 11.38%. The density achieved was as low as 35.2 kg/m^3 (2.2 pcf).

8.7.4.2   Effect of Blowing Agent Concentration

Figures 8.8 and 8.9 illustrate that the foam density decreases with increase in the level of blowing agent. This result is expected because the presence of more blowing agent expands the polymer more and thus yields lower density. On the other hand, the foam growth time decreases with increase in the concentration of blowing agent. Similar results have been reported earlier in the literature [273]. Higher blowing agent concentration leads to accelerated diffusion rate. This allows bubbles to grow quickly and thus reach final (equilibrium) radius at an earlier time. The predictions of the model with the new gas escape boundary condition seem to agree well with the experimental data. The maximum deviation observed in the predicted final foam density when compared with the experimental data is about 8%.

8.7.5   Conclusions

- A successful and reliable process for producing environmentally friendly and water-soluble biodegradable polyvinyl alcohol foam can be predicted by foam growth theory and achieved by experimental trials. The lowest foam density achieved was about 35 kg/m^3 (2.2 pcf).

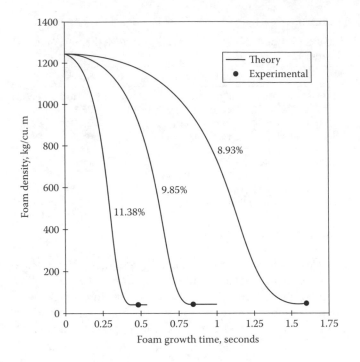

FIGURE 8.8
Foam Density as a Function of Bubble Growth Time for Various Blowing Agent Concentrations.

- Use of cross-linking agent is important to achieve low density foam.
- The predictions of the heat transfer coupled foam growth model seem to agree with the experimental data well.

8.8   Starch Foams

Agricultural materials like starch are very attractive for the foaming process. Starch is a renewable resource polymer available for plenty of commercial uses. Several consumer reports praise the use of starch as an alternative material to compete against polystyrene in the loose-fill market. The low cost nature of polystyrene based loose-fill products is still attractive to some but many are looking for environmentally better material. It is good to review the status of starch foams and cover the latest developments in the field. Experts conservatively believe that the starch-based loose-fill market size to be about 10% of the polystyrene loose-fill market [276]. Tatarka [276] gave an excellent review of starch foam technology in 1995 and claimed that several companies actively developed starch-based loose-fill prod-ucts—American Excelsior, Free-flow Packaging, Clean Green Packaging,

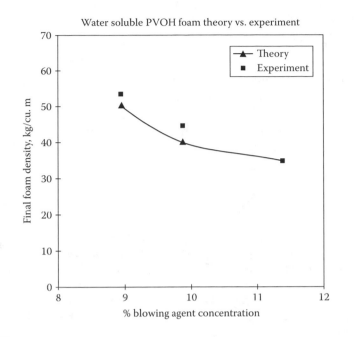

FIGURE 8.9
Comparison of Theory with the Experimental Data.

Storopack and EnPac. There are more companies involved in this research and foaming work in the past two decades.

8.8.1 Chemistry of Starch

Starch is a polysaccharide polymer made up of repeating glucose groups linked by glycoside linkages. The chain length varies with plant source and origin, and the average length generally ranges between 500–2000 repeat units. Starch can be classified into two major molecules based its chemical structure—amylose and amylopectin. The amylose starch is flexible due to its alpha linkage. Amylose represents the linear chain molecules and amylopectin represents highly branched molecules. In order to make good foams, we need some amylose content to keep the foam flexible and some amylopectin content to have good melt strength and foamability. There are several types of starches available. A few examples are shown below.

8.8.2 Types of Starches

- Corn starch
- Pregelatinized corn starch
- Dextrins

- Tapioca starch
- Potato starch
- Native starch
- Hydroxy-ethylated starch
- Hydroxy-propylated starch

Typical price range for some of the above starches is given in a paper presented by Tatarka [276] in 1995, but the prices may vary significantly today.

8.8.3 Extrusion of Starch Foams

Production of starch-based foam can be derived from extrusion cooking technology that dates back more than 60 years. This process simply means melting of starch material and then mixing it with a blowing agent such as water and then forcing the molten mass through a die to cause foam expansion. Nucleation of cells occurs due to sudden pressure drop and then porous structure results due to diffusion of steam into nucleated bubbles causing rapid expansion. During extrusion, the screw geometry, pressure, cooling ability, moisture content, type of starch, chemical additives, nucleating agents, temperature distribution and control seem to play a big role in controlling the foam density and properties. All types of extruders can be used although twin-screw extruders are preferred due to good mixing and cooling to maintain homogeneity. It also allows for a short residence time inside the extruder to avoid overcooking or degradation. The major advantage of the extrusion process is low-cost continuous operation and easy injection of single or multiple blowing agents to achieve desired properties. Figure 8.10 shows its cell structure with 25 times magnification. The challenge is to eliminate corrugation and increase the percentage of closed cells. A lot more research is needed to accomplish these goals. A typical starch-based automotive packaging product is shown in Figure 8.11 [266].

Another method is pellet expansion where the polymer pellets are soaked into a low boiling point fluid and then rapidly heated up in a closed mold. This method is used to produce a molded part.

8.8.4 Starch Foam Compositions

This section covers a description of extrusion of starch foams using various types of starches and additives done by Ramesh. The following table shows the list of materials used and their functions. The foaming process is very similar to the set up shown in Figure 8.6 and the experimental procedure is identical. Basically the following additives were mixed and added through the feed section of a co-rotating twin-screw extruder and blowing agent mixture (methanol and water) were added. The molten polymer charged

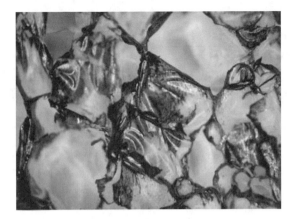

FIGURE 8.10
Cell Structure.

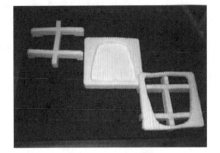

FIGURE 8.11
Typical automotive packaging using starch foam.

with blowing agent was cooled and stabilized before being sent to the die. A slit die having a width and opening of 1.3 cm and 0.16 cm was used for extrusion of starch foam ribbons. Mixtures of various compositions are reported in Table 8.7. The details of foaming temperature and die pressure are shown in Tables 8.8 to 8.10 along with foam properties.

In addition to the above ingredients, polyethylene glycol, Sorbitol, Glycerol monosterate and polyethylene oxide were individually added to the standard formulation to study their plasticizing effects. The goal is to produce a strong but flexible and resilient starch foam. Although the above ingredients seem to help to some extent, it is not easy to keep starch foam away from brittleness especially when the foam is exposed to a very low humidity environment for a lengthy period of time. The following table shows the formulation of extrusion runs of Starch A, B and C mentioned in Table 8.7.

TABLE 8.7

Common Ingredients for Starch Foam Extrusion

Ingredients	Type/ Functional Role	Composition	Purpose
Starch A	Hydroxypropylated starch	23% amylose and 77% amylopectin	Hydropropyxylation is expected to lower gelatinization
Starch B	Native Wheat Starch	23% amylose and 77% amylopectin	Same as above
Starch C	High Amylose Corn Starch	50% amylose and 50% amylopectin	More flexible due to higher amylose content
EVOH	Compatibilizer Add crystallinity	38% and 44% PE content	Additive to open cells to achieve lower density
PVOH	Strengthening agent	88% PVOH content and some glycerol	Water solubility, adds more flexibility and rigidity
Safoam	Nucleating agent	Decompose into gases and nucleate cells	Create fine cells and achieve lower foam density
Silicon dioxide	Fine particles Nucleating agent Feed assisting agent	Approx. 0.007 mm in size	Avoid bridging the feed hopper for smooth extrusion
HIPS	Enhance flexibility	Rubber particles in PS matrix	Ultra low foam density (20 kg/m^3)
Methanol and water	Blowing agents	Plasticization and foam expansion	Achieve lighter weight foam

TABLE 8.8

Extrusion Foaming Runs with High Amylose and Hydroxypropylated Starch Mixture

Ingredients	Parts by Weight	Foam Properties and Processing Conditions
Starch A	40	Low density
Starch C	20	Good foam
EVOH	18	Slightly brittle
HIPS	10	Die Pressure = 669 psi
Safoam	1.5	Foaming Temperature = 130 F
Silica	0.5	Foam Density = 1.27 pcf
Methanol	7.0	Moisture sensitive
Water	3.0	Soluble in water

TABLE 8.9

Extrusion Foaming Runs with Native Starch and High Amylose Starch Blend

Ingredients	Parts by Weight	Foam Properties and Processing Conditions
Starch B	20	Low density
Starch C	40	Good foam
EVOH	30	Flexible, but becomes brittle at low RH
HIPS	10	Die Pressure = 1050 psi
Safoam	1.5	Foaming Temp = 133 F
Silica	0.5	Foam Density = 1.83 pcf
Methanol	4.5	Prevented Cell Collapsing
Water	6.0	Soluble in Water

TABLE 8.10

Extrusion Foaming Runs with High Amylose Starch

Ingredients	Parts by Weight	Foam Properties and Processing Conditions
Starch C	60	Flexible, but becomes brittle at low RH
EVOH	30	Die Pressure = 992 psi
Safoam	1.5	Foaming Temp = 132 F
Silica	0.5	Foam Density = 1.30 pcf
Methanol	4.5	Prevented Cell Collapsing
Water	5.0	Soluble in Water

8.8.5 Experimental Results and Discussion

The following is a brief summary of various starch blend foams.

8.8.5.1 *High Amylose and Hydroxypropylated Starch Blend Foam*

Foams made with high amylose starch exhibit low density (1.27 lb/ft^3 or 20 kg/m^3) and some flexibility. The presence of methanol helped to reduce foam collapsing due to condensation of water inside the cells that causes partial vacuum responsible for shrinking of extruded foam. Slight brittleness may be due to the use of HIPS and EVOH additives. More work needs to be done in the future to optimize flexible starch foam formulations.

8.8.5.2 *Native Starch and High Amylose Starch Blend Foam*

A good quality foam having 1.83 lb/ft^3 or 29.3 kg/m^3 density can be made as shown in Figure 8.11. The foam seems to show good flexibility and good cell structure. However, when the foam is exposed to low relative humidity environment for a prolonged period of time, it becomes slightly brittle.

8.8.5.3   *High Amylose Starch Blend Foam*

The use of Starch C (50/50 amylose/amylopectin ratio) seems to yield excellent soft and flexible foam with 1.3 lb/ft³ or 20.8 kg/m³ density foam. It was observed that the high amylose starch based foam tend to expand more in the machine direction. On the other hand, low amylose starches seem to expand in the radial direction more as the extrudate foam exits the die. Addition of 3% polyethylene glycol or 10% Vinex 2034 (PVOH) with high glycerine and 2.5% sorbitol individually gave higher density foams closer to 4 lb/ft³ or 64 kg/m³. This clearly shows the sensitivity of the process to various additives and one must optimize the process for customer requirements and desired features. Once again this foam becomes brittle when exposed to RH less than 15% for a long period of time.

8.8.6   Key Observations

1. High amylose content foams gave excellent flexibility, but it needs to be improved further to perform at low humidity conditions.
2. PVOH did not serve well as a useful additive as expected.
3. Addition of HIPS lowered the foam density and enhanced flexibility and toughness.
4. The slit die seems to work well for foam extrusion. Higher output rate would be helpful to create additional die pressure necessary to lower the density even further.
5. Methanol seems to act as the best blowing agent.

There are several scientists including Ramesh and Malwitz [269] who worked on extrusion of flexible biodegradable foams. In addition to prior excellent technical accomplishments, Narayan [277–279] has submitted several technical papers recently on the extrusion of starch foams. The details can be found in his future publications. It was found that screw configurations are important to control the foaming process. Figure 8.12 shows a SEM micrograph of starch foam. Typically starch foam surfaces are rough and abrasive, however by using biodegradable thermoplastic polymer additives, the surface can be modified to provide a smooth finish [278].

8.8.7   Comparison of Starch Foam vs. PS Foam

Tatarka [276] reported the properties of commercial starch-based loose-fill products compared to those of recycled polystyrene loose-fill. For comparison, a small example is given in Table 8.11.

The data in this table shows clearly that the bulk densities of starch foams are about double the value of polystyrene loose-fill foams. One of the reasons

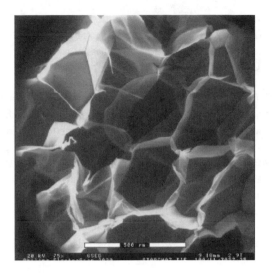

FIGURE 8.12
SEM Picture of starch foam.

TABLE 8.11

Comparison of Starch Foam vs. Polystyrene Foam [277]

Properties	Starch Foam	Polystyrene
Bulk density, kg/m^3 (pcf)	9.61 (0.6)	4.16 (0.26)
Price, \$/m^3 (\$/ft^3) (1995)	21.89 (0.62)	13.42 (0.38)
Cell size, mm	0.6–0.8	0.15–0.35
Cell Geometry	Irregular, more open cells	Regular, pentagonal

is the basic difference in the non-foamed density of the polymers. Starch material is about 30% denser than polystyrene to start with and therefore a higher foaming efficiency is needed to reduce the density to match with PS. Starch-based loose-fill is also more expensive than recycled polystyrene and hence lower density is desired. Another reason is large cell size and poor cell structure. In order for starch to foam well it must have good extensional viscosity properties and improved melt strength. Polystyrene is an amorphous polymer and very easy to foam because it has a wide operating window and automatically gives pentagonal shaped closed cells. On the other hand, it is very hard to make close cell starch foam below 30 kg/m^3 due to poor melt strength, narrow processing window and moisture sensitivity.

It is well known that the cell morphology plays a vital role in controlling mechanical properties. PS cell size is smaller and geometry is regular with good polymer distribution in the cell wall, junction and struts, whereas the cell structure of starch foam is poor with irregular cell walls and more open cells. More improvement is needed from the technical and equipment side to make starch foams viable for commercial applications. The future of starch foam looks bright as scientists are developing new process aids and dies for foam extrusion. One of the key needs is to make a starch foam to perform for packaging application that require elevated temperatures in a humid environment. There are environmental and safety issues such as dusting at low humidity and shrinkage particularly at relative humidity higher than 50%. There are several ways to improve foam structure by blending other resource renewable polymers. Increasing the output rate, changing the die geometry, improving packaging design and decreasing foam density are several important steps to achieving highest performance at lowest possible cost. Once the cost becomes low then it is easy to compete against polystyrene and other polymeric materials since the resin prices are rising sharply. More information on starch foams can be found from Guan, Hanna [280] and Fang, Hanna [281].

8.9   Polylactic Acid (PLA) Polymer

Polylactic acid polymer is a very good biodegradable & renewable resource polymer for foaming applications. It falls under the category of biodegradable polyesters and is derived from corn. Polyesters represent a broad class of polymers with a subset that are degradable due to either ester bonds hydrolysable under degradation conditions (e.g., PLA) or direct attack of microorganisms (e.g., PHAs). Polyesters can be classified into two major groups as shown below. Some of the most important polymers are mentioned below.

Aliphatic Polyesters (ALP)
  • Polylactic Acid (PLA)
  • Polyhydroxyalkanoates (PHA)
  • Polycaprolactone (PCL)
  • Polyhydroxybutyrate (PHB)
  • Polyhydroxyhexanonate (PHH)
Aromatic Polyesters (ARP)
  • Polyethylene terephthalate (PET)
  • Aliphatic-aromatic copolyesters (AAC)
  • Polymethylene adipate/terephthalate (PTMAT)

The heat resistance of parts made from PLA is determined by the amount of crystallinity in the parts. In its amorphous state, PLA lacks high temperature stability due to its relatively low glass transition temperature. This can be overcome by developing crystallinity in either the extruded foam or thermoformed part. In addition, the toughness of PLA can be improved significantly by orientation of the amorphous phase which is easy to achieve in the foaming process. Without this orientation, PLA lacks the toughness required in many applications. On the other hand, aromatic polyesters such as PET generally exhibit high heat resistance and excellent resistance to chemical and moisture attack. PET foam is also good for thermal insulation. Aliphatic polyesters are synthesized either from diols and dicarboxylic acids through condensation polymerization reaction or from ring opening polymerization (ROP) of lactones. It is possible to blend these polyesters with starches mentioned earlier to achieve lower cost and improve moisture resistance of starch polymers.

8.9.1 Chemistry of Polylactic Acid (PLA) Polyesters

Polylactic Acid (PLA) is a linear polymer available in amorphous and semi-crystalline forms. This polymer can be foamed to reduce its weight to serve in packaging and non-packaging applications. PLA polymer is biodegradable and has good mechanical and flexible properties. With the rising cost of petroleum based resins, it is important to look at the renewable resource polymers closely. However, foaming of PLA is difficult due to poor melt strength and other issues like surface roughness and poor cell structure. Chain-extender additives can be used to improve the melt strength of PLA.

Chemically, PLA is a linear thermoplastic polymer formed by ring opening polymerization of lactide. Lactic acid is the starting material for lactide and is produced through fermentation of natural sugars from a readily available source such as corn. Lactic acid is converted into lactide and then into PLA (poly (3,6-dimethyl-1,4-dioxan-2,5-dione). Up to 3 lactide monomers are involved: meso-Lactide, D-lactide and L-lactide. D and L type of lactides melt at higher temperatures than meso-Lactide. The amount of D- or meso-lactide present in the L-PLA polymer changes the properties significantly in terms of melting temperature, crystallization rate and therefore processibility and properties of foams. For example, the higher the D-isomer content in the polymer then the lower the crystallization rate and lower the melting point. NatureWorks, LLC is the major producer of PLA polymer in the world, however several other smaller producers are entering the market. PLA has been certified to be fully compostable by DIN CERTCO (E DIN 54900-1, -2 and -3). With opening of a world scale production plant for PLA in 2001, there has been a significant increase in the amount of development work on PLA foam. Recently Hu and team members investigated the solubility and diffusion of carbon dioxide in PLA at room temperature and pressures up to 5.8 MPA [282]. So there seems to be a lot of interest among academic and industrial scientists and engineers.

8.9.2   PLA Foam

PLA foam research is evolving. There is so much more to learn in understanding and processing PLA materials. Like any other foam process, a typical PLA foam extrusion process has three important steps: extrusion, stabilizing and winding. One difference with working with PLA rather than conventional resins is that moisture can degrade the PLA resin. Therefore, PLA must be dried prior to extrusion. Foaming can be made in various types of extruders. PLA polymer is typically mixed with nucleating agents such as talc and other process aids to improve foaming and cell formation. Typical blowing agents used are HCFC, carbon dioxide and hydrocarbons. While current assets used for polystyrene can be used to produce PLA foams, modifications to screw designs (both primary and secondary) and take off equipment will result in maximized output and the best foam quality. Because of the low T_g of the polymer, screw designs that maximize cooling and minimize shear heating are desired in the cooling extruder. PLA has been extruded into foams at higher densities and lowering the density with good thermal stability is still a challenge. As pointed out earlier, the ratio of d to l-lactide isomer in the resin affects the degree of crystallinity. Lower levels of d-lactide in the resin have the potential to produce foams with high levels of crystallinity and dimensional stability up to 140°C, while resins with high levels of d-isomer will result in a completely amorphous foam that has stability only up to 45°C. Figure 8.13 shows a typical example of an extruded PLA foam.

Recent work done by Hu and co-workers showed that PLA samples produced very nice foams when saturated at 2.8 MPa and 4.1 MPa with carbon dioxide for 2 days using a batch microcellular foaming process followed by

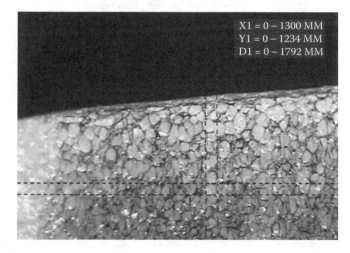

FIGURE 8.13
PLA rod foam micrograph made by Natureworks with cell size of 130 micron.

foaming at 373 K [283]. The cell diameters were as fine as 30–40 micrometers and cell densities of about 7.93×10^7 cells/cm^3. However, there seems to be a huge gap between batch microcellular process and continuous extrusion process. A technical breakthrough is required to narrow the gap and reduce the slope of the steep learning curve. More information on biodegradable microcellular foams can be found from Fujimoto et al. [282]. More information on PLA foams can be found from Huneault [284], Guan et al. [285], Di et al. [286], Ray et al. [287], and Lu et al. [288].

8.9.3 PLA Cellular Structure through Solution Process—Another Method

It is well known that PLA foam can be made using the extrusion foaming process as we described before. However, there is some research effort going on in the area of solution process. Porous cellular structures of PLA through phase separation in its solution have been reported by several researchers [289,290]. Nam et al. reported thermally induced phase separation technique and Hua et al. used water to induce phase separation of PLA solution in 1.4-dioxane, showing pore size influenced by the annealing time after gel formed. Recently, PLA gel in acetone was prepared by Zhang et al. [43] using a temperature induced phase separation technique and compared to the gel prepared in tetrahydrofuran. Interesting crystal structures and morphologies were noticed. The purpose of these studies was to study the phase transitions in PLA systems and to focus on its crystallization effects and morphological structures. These fundamental studies are expected to bring new insights into developing foamable PLA materials in the future.

8.10 Future Work

There is a lot of focus on biodegradable foam research and much more needs to be accomplished by future scientists and engineers. Recently, Iannace and co-workers presented some interesting work on biodegradable foams based on poly-caprolactone (PCL) and poly-lactic acid (PLA) materials [291]. They demonstrated how to improve the foaming process through reactive processing and nanocomposites. Still a breakthrough is needed to make thick and flexible starch foam at lower cost. Further, the effect of D-content on foamability should be further explored for advancing the technical knowledge. Dusting and brittleness issues must be solved to customer satisfaction. Recent developments on biodegradable polyesters are interesting. Use of wood fiber for foaming application is increasing [292]. Similarly, there are some recent developments in biodegradable film research including cellulose coatings and agro-polymers for active packaging applications. The more we get into this research, the easier it is to understand that the research field is wide open and very broad. So a careful selection of biodegradable materials and foaming processes and equipments is necessary to contribute to society and make a difference in the world.

8.11 Nomenclature

c	=	Gas concentration
D	=	Diffusion coefficient
E_V	=	Activation energy for viscosity equation
E_D	=	Activation energy for diffusion equation
ΔH_s	=	Heat of solution
l	=	Foam sheet thickness
R_g	=	Gas constant
r	=	Radial Coordinate
x_c	=	Degree of crystallinity
y	=	Mole fraction of blowing agent
α	=	Thermal diffusivity
η_o^*	=	Zero-shear-viscosity according to equation (8.2)
ε/κ	=	Lennard-Jones temperature

References

1. Weaire, D. and Hutzler, S., *The Physics of Foams*, Oxford University Press, New York, 1999, chap. 1.
2. Perkowitz, S., *Universal Foam*, Anchor Books, New York, 2001.
3. Benning, C.J., *Plastic Foams, Vol. I & II*, John Wiley & Sons, New York, 1969.
4. Stralen, S.V. and Cole, R., *Boiling Phenomena, Vol. 1*, Hemi-Sphere/McGraw-Hill Co., New York, 1979.
5. Lee, S.T., *Foam Extrusion*, Lee, S.T., Ed., CRC Press (ex Technomic), 2000, chap. 4.
6. Sadoc, J.F. and Rivier, N., Ed., *Foams and Emulsions*, Kluwer, Dordrecht, The Netherlands, 1999.
7. Biesenberger, J.A., *Devolitilization of Polymers*, Hanser, Munich, 1983, chapter 1.
8. Rubens, L.C., *Perspectives on Cellular Materials; From Synthetic Foam Cushions to HumanLungs*, 52nd Annual Tech. Conf., Sopn. Soc. Plas. Eng., preprint 1934–1940, San Francisco, CA, 1994.
9. Biesenberger, J.A. and Sebastian, D.H., *Principles of Polymerization Engineering*, John Wiley & Sons, New York, 1983, chap. 6.
10. Buist, J.M. and Gudgeon, H., *Advances in Polyurethane Technology*, John Wiley and Sons, New York, 1968.
11. Klempner, D. and Frisch, K.C., Ed., *Polymeric Foams*, Hanser, Munich, 1991.
12. Throne, J.L., *Thermoplastic Foams*, Sherwood, Hinckley, OH, 1996.
13. Lee, S.T., *Foam Extrusion*, Lee, S.T., Ed., CRC Press (ex Technomic), 2000, chap. 1.
14. Suh, N.P., *Innovation in Polymer Processing*, Stevenson, J., Ed., Hansen, 1996, chap. 3.
15. Lee, S.T. and Park, C.B., *Thermoplastic Foam Course*, Technomic, Lancaster, U.K., 2000.
16. Woods, G., The ICI Polyurethyanes Book, ICI Polyurethancs, 1990.
17. Herrington, R., Dow Polyurethanes; Flexible Foams, Dow Chemical, 1997.
18. Munters, G. and Tandberg, J.G., U.S. Patent 2,023,204, 1935.
19. Frisch, K.C., Historical Developments of Polyurethanes, *60 Years of Polyurethanes*, Kresta and Eldred, Ed., Technomic, 1998.
20. Johnson, F.L., U.S. Patent 2,256,483, 1941.
21. Kennedy, R.N., *Handbook of Foamed Plastics*, Benders, R.J. Ed., Lake, Libertyville, IL, 1965, Section XII.
22. Bayer, O., Rinke, H., Siefken, W., Ortner, L. and Schild, H., German Patent 728, 981, 1942.
23. Stastny, F. and Gaeth, R., U.S. Patent 2,681,321, 1954.
24. Beyer, C.E. and Dahl, R.B., U.S. Patent 3,058,161, 1962.
25. Rubens, L.C., Griffin, J.D., Urchick, D., U.S. Patent 3,067,147, 1962.
26. Spa, L.M.P., Italian Patent 795,363, 1967, also British Patent 1,152,306, 1969.
27. Woollard, D.C., *J. Cellular Plastics*, 4, 1, 16–21, 1968.
28. Parrish, R.G., U.S. Patent 3,637,458, 1972.

29. Collins, F., U.S. Patent 4,323,529, 1982.
30. Hirosawa, K. and Shimada, S., U.S. Patent 4,379,859, 1983.
31. Xanthos, M. and Dey, S., *Foam Extrusion*, Lee, S.T., Ed., CRC Press (ex Technomic), 2000, chap. 11.
32. Gibson, L.J. and Ashby, M.F., *Cellular Solids*, Pergamon, Oxford, London, 1988.
33. Van Krevelin, D.W., *Properties of Polymers*, 3rd ed. Elsevier, New York, 1990, chap. 13, 18.
34. Kolossow, K.D., *Plastics Extrusion Technology*, Henson, F., Ed., Hanser, Munich, 1988, chap. 13.
35. Martelli, F., *Twin Screw Extruders*, Van Nostrand Reinhold Co., New York, 1983, chap. 7.
36. Glicksman, L., *Foams and Cellular Materials: Thermal and Mechanical Propertes*, Massachussetts Institute of Technology, summer course notes, Cambridge, MA, 1992.
37. Midgley, T., Henne, A.L. and McNary, R.R., *Heat Transfer*, U.S. Patent 1,833,847.
38. Midgley, T., Henne, A.L. and McNary, R.R., *Manufacture of Alphatic Fluoro Compounds*, U.S. Patent 1,930,129, 1933.
39. Wei, J., *The Third Paradigm of Chemical Engineering: Molecular Product Engineering*, seminar, New Jersey Institute of Technology, Apr., 2001.
40. Williams, D. and Bogdan, M., *HFC-245fa & HFC-245fra blends: Blowing Agent Solutions for All Rigid and Integral Skin Foam Applications*, Foam Conference, RAPRA, Frankfurt, Germany, 2001.
41. Zipfel, L. and Dournel, P., *Optimization of Insulation Performance by HFC 365mfc*, Foam Conference, RAPRA, Frankfurt, Germany, 2001.
42. Freedonia Group Report, *Foamed Plastics*, Freedonia Group Inc., Cleveland, OH, 2002.
43. Zhang, J., *Phase transition of Poly(l-lactic acid) in Bulk and in Solutions*, Ph.D. thesis, Chem. Eng. Dept., New Jersey Institute of Technology, 2005.
44. Himmelblau, D.M., *Basic Principles and Calculations in Chemical Engineering*, 3rd ed., Prentice-Hall, 1974, chap. 7.
45. Handa, P. and Zhang, Z., *Sorption of CO_2 in Polycarbonates*, 7th FoamTech Meeting, Canadian National Research Council, Jun., 1999.
46. Florry, P. J., Thermodynamics of High Polymer Solutions, *J. Chem. Phys.*, 10, 51–61, 1942.
47. Huggins, M.L., Theories of Solutions of High Polymers, *J. Amer. Chem. Sci.*, 64, 1712–1720, 1942.
48. Biesenberger, J.A., *Devolatilization of Polymers*, J.A. Biesenberger, Ed., Hanser, Munich, 1983, chap. 1.
49. Gorski, R.A., Ramsey, R.B. and Dishart, K.T., Proc. SPI 29th Ann. Tech. Mark. Conf., 1983.
50. Lee, S.T., A Fundamental Study of Thermoplastic Foam Extrusion with Physical Blowing Agents, in *Polymeric Foams*, Khemani, K.C., Ed., ACS Symposium Series 669, 1997, chap. 13.
51. Chaudary, B.I. and Johns, A.I., Solubilities of Nitrogen, Isobutane and Carbon Dioxide in Polyethylene, *J. Cell. Plast.*, 34, 312–328, 1998.
52. Stiel, L.I. and Harnish, D.F., Solubility of Gases and Liquids in Molten Polystyrene, AIChE, 22, 117, 1976.
53. Nawaby, A.V. and Zhang, Z., Solubility and Diffusivity, in *Thermoplastic Foam Processing*, Gendron, R., Ed., CRC Press, Boca Raton, FL, 2004, chap. 1.

54. Sato, Y., Wang, M., Takishima, S., Masuoka, H., Watanabe, T. and Fukasawa, Y., Solubility of Butane and Isobutane in Molten Polypropylene and Polystyrene, *Poly. Eng. Sci.*, 44, 11, 2083–2089, 2004.
55. Sanchez, I.C. and Lacombe, R.H., Statistical Thermodynamics of Polymer Solutions, *Macromolecules*, 11, 6, 1145–1156, 1978.
56. Kiszka, M.B., Meilchen, M.A., and McHugh, M.A., Modeling High-Pressure Gas-Polymer Mixtures Using the Sanchez-Lacome Equation of State, *J. App. Poly. Sci.*, 36, 583–97, 1988.
57. Garg A., Gulari, E. and Manke C.W., Thermodynamics of Polymer Melts Swollen with Supercritical Gases, *Macromolecules*, 27, 5643–5653, 1994.
58. Biesenberger, J.A. and Lee, S.T., A Fundamental Study of Polymer Melt Devolatilization: II, A Theory for Foam-Enhanced DV, 44th *SPE ANTEC* preprint, 846, 1986.
59. Martini, J.F., Master's thesis, Mech. Engr. Dept., MIT, 1981.
60. Dankwert, P.V., Chem. Eng. Sci., 2, 1, 1953.
61. Pontiff, T., Foaming Agents for Foam Extrusion, in *Foam Extrusion*, Lee, S.T., Ed., CRC Press (ex Technomic), 2000, chap. 10.
62. Eckardt, H., *Low Pressure Injection Moulding*, Foaming Conference RAPRA, Frankfurt, Germany, 2001.
63. Michaeli, W., Pfannschmidt, O. and Habibi-Naini, S., Injection Moulding of Microcellular Foams, *Kunststoffe*, 92, 8, 56–60, 2002.
64. Miyamoto, A., Akiyama, H. and Usunda, Y., Process for the Preparation of Closed-Cellular Shaped Products of Olefin Polymers Using a Mixture of Citric Acid Salt and a Carbonate or Bicarbonate as the Nucleating Agent, U.S. Patent 3,808,300, 1974.
65. Kiyono, H., Ishimoto, A., Noda, Y. and Yada, K., Apparatus for Continuously Producing Foamed Sheets, U.S. Pat. 4,124,344, 1978.
66. Gendron, R. and Daigneault, L.E., Rheology of Thermoplastic Foam Extrusion Process, in *Foam Extrusion*, Lee, S. T., Ed. CRC Press (ex Technomic), 2000, chap. 10.
67. Simha, R. and Moulinie, P., Statistical Thermodynamics of Gas Solubility in Polymers, in *Foam Extrusion*, Lee, S.T., Ed., CRC Press (ex Technomic), 2000, chap. 2.
68. Utraki, L.A. and Simha, R., Free Volume and Viscosity of Polymer-Compressed Gas Mixtures during Extrusion Foaming, *J. Polym. Sci. B: Polym. Phys.*, 39, 342–362, 2001
69. Fridman, M.L., Sabsai, O.Y., Nikolaeva, N.E. and Barshtein, G.R., Rheological Properties of Gas-Containing Thermoplastic Materials during Extrusion, *J. Cell. Plas.*, 25, 574–595, 1989.
70. Todd, D. and Subir, D., *Visulization of Foam Processing*, video in Polym. Proc. Inst. at New Jersey Institute of Technology, 1998.
71. Ramesh, N.S., Lee, S.T. and Lee, K., Novel Method for Measuring the Extensional Viscosity of PE with Blowing Agent and Its Impact on Foams, Foams 2002, *SPE*, 113–120, Houston, 2002.
72. Dey, S.K., Jacob, C., and Biesenberger, J.A., Effect of Physical Blowing Agents on Crystallization Temperature of Polymer Melts, *Annu. Tech. (ANTEC) Conf. Proc.*, 2197–2198, Soc. Plas. Eng., 1994.
73. Avrami, M., Kinetics of Phase Change. I, *J. Chem. Phys.*, 1103, 1939.
74. Khanna, Y.P. and Taylor, T.J., *Poly. Engr., Sci.*, 28, 1042, 1988.
75. Keller, A., An Approach to Phase Behavior, *Macromol. Symp.*, 98, 1–42, 1995.

76. Yamaguchi, M., *Polymeric Foams: Mechanisms and Materials*, ed. Lee and Ramesh, Pub. CRC Press, Boca Raton, FL, 2004, chap. 2.

77. Park, C.B., Ph.D. thesis, Mech. Engr. Dept., Massachussetts Institute of Technology, 1993.

78. Burnham, T.A., Cha, S.W., Kim, R.Y., Anderson, J.R., Stevenson, J.F., Suh, N.P. and Pallaver, M., Method and Apparatus for Polymer FoamExtrusion, In Particular Microcellular Foam, EP Patent 923,443, 2002.

79. Connahan, V.D., Ph.D. thesis, Chem. Engr. Dept., University of Maryland, 1971.

80. Kabayama, M.A., Long-Term Thermal Resistance Values of Cellular Plastic Insulations, *J. Ther. Insu.*, 10, 286–300, 1987.

81. Brandrup, J., Immergut, E.H. and Grulke, E.A., *Polymer Handbook*, 4th edition, John Wiley and Sons, 1999.

82. Norton, F.J., Thermal Conductivity and Life of Polymer Foams, *J. Cellu. Plas.*, 23–37, 1967.

83. Park, C.P., Expandable Polyolefin Compositions and Preparation Process Utilizing Isobutane Blowing Agent, U.S. Patent 4,640,933, 1987.

84. Yang, C.T. and Lee, S.T., Dimensional Stability Analysis of Foams Based on LDPE and Ethylene-Styrene Interpolymer Blends, 60th *SPE ANTEC* preprints 1782–1786, 2002.

85. Cronin, E. W., Extrusion Process for Polyolefin Foam, U.S. Patent 3,644,230, 1972.

86. Yang, C.T. and Lee, S.T., *Dimensional Stability Analysis for Polyethylene*, Frados, J., Ed., *Plastics Engineering Handbook*, 4th ed., Society of the Plastics Industry, Van Nostrand Reinhold, New York, 1976.

87. *Polyolefin Foams 1996-2001, North America, Europe & Japan*, Chemical Market Resources, Inc., Houston, Texas, 1997.

88. Eaves, D., *Polymer Foams*, RAPRA Tech. Ltd., 2001, chap. 2.

89. Yasunaga, K., Neff, R.A., Zhang, X.D. and Macosko, C.W., Study of Cell Opening in Flexible Polyurethane Foam, *J. Cellu. Plas.*, 32, 427–448, 1996.

90. Elwell, M.J., Mortimer, S. and Ryan, A.J., *Macromolecules*, 27, 5428, 1994.

91. Parrish, R.G., Process for the Preparation of Low Density Microcellular Foam Sheets Exhibiting High Work-to-Tear Values, US Patent 3,787,543, 1974.

92. Kitamori, Y., PE-PS Alloy Foam, Proc. Thermoplastic Foams Technical Conference, Industrial Technology Research Institute, Taipei, Taiwan, 1995.

93. Ryan, A.J., Willkomm, W.R., Bergstrom, T.B., Macosko, C.W., Koberstein, J.T., Yu, C.C. and Russel, T.P., Dynamics of (Micro)Phase Separation During Fast, Bulk Copolimerisation: Some Synchrotron Small-Angle X-Ray Scattering Experiments (On Bulk Formation of a Copolyurethane), *Macrom.*, 24, 10, 2883–2889, 1991.

94. Gedde, U.W., Crystallization and Morphology of Binary Blends of Linear and Branched Polyethylene, *Progr. Colloid Polym. Sci.*, 87, 8–15, 1992.

95. Utracki, L.A., Foaming Polymer Blend, FoamTech, Canadian National Research Council, Dec., 1998.

96. Yoshimura, S. and Kuwabara, H., Process for Producing Prefoamed Polymer Particles, U.S. Patent 4,464,484, 1984.

97. Sasaki, H., Sakaguchi, M., Akiyama, M. and Tokoro, H., Expanded Polypropylene Resin Beads and A Molded Article, U.S. Patent 6,313,184, 2001.

98. Suh, N.P., *Microcellular Plastics*, in *Innovation in Polymer Processing*, Sterenson, J.F., Ed., Hanser, 1996, chap. 3.

99. Lee, S.T., A Fundamental Study of Foam Devolatilization, in *Polymer Devolatilization*, Albalak, R.J., Ed., Marcel Dekker, Inc., New York, 1996, chap. 6.
100. Blander, M. and Katz, J.L., Bubble Nucleation in Liquids, *AIChE J.*, 21, 5, 833–848, 1975.
101. Kumar, V., Ph.D. thesis, Mech. Eng. Dept., Massachussetts Institute of Technology, 1988.
102. Colton, J.S. and Suh, N.P., The Nucleation of Microcellular Foam with Additives, I: Theoretical Considerations, *Polym. Eng. Sci.*, 27, 485–503, 1987.
103. Han, J.H. and Han, C.D., Bubble Nucleation in Polymeric Liquids, I: Bubble Nucleation in Concentrated Polymer Solutions, *J. Polym. Sci. B: Polym. Phys.*, 28, 711, 1990.
104. Ruengphrathuengsuka, W., Bubble Nucleation and Growth Dynamics in Polymer Melts, Ph.D. thesis, Chem. Eng. Dept., Texas A&M University, 1992.
105. Han, J.H. and Han, C.D., Bubble Nucleation in Polymeric Liquids II. Theoretical Considerations, *J. Polym. Sci., B: Polym. Phys.*, 28, 743–761, 1990.
106. Reid, R.C., Prausnitz, and Poling, B.E., *The Properties of Gases and Liquids*, 4th ed., , McGraw-Hill, New York, 1986, p. 644.
107. Taki, K., Yatsuzuka, T., Hirano, T. and Ohshima, M., Visual Observation of Polymeric Foaming with Super Critical fluid, PPS18, Paper 341, Guimarães, Portugal, 2002.
108. Taki, K., Yatsuzuka, T., Nakayama, T. and Ohshima, M., Visual Observation of Batch & Continuous Foaming Processes, Foams 2002, SPE, 141–151, Houston, 2002.
109. Lee, S.T., *Volatile Blowing Agent Considerations for Foam Extrusion*, Foams Retec, New Jersey, 1999.
110. Chen, L., Wang, X., Straff, R. and Blizard, K., Shear Stress Nucleation in Microcellular Foaming Process, *Polym. Eng. Sci.*, 42, 6, 1151–1158, 2002.
111. Amon, M. and Denson, C.D., A Study of the Dynamics of Foam Growth: Analysis of the Growth of Closely Spaced Spherical Bubbles, *Polym. Eng. Sci.*, 24, 13, 1026–1034, 1984.
112. Ramesh, N.S., Rasmussen, D.H., and Campbell, G.A., Numerical and Experimental Studies of Bubble Growth During the Microcellular Foaming Process, *Polym. Eng. Sci.*, 31, 1999.
113. Lee, S.T., Ph.D. thesis, Chem. Engr. Dept., Stevens Institute of Technology, 1986.
114. Lee, S.T. and Ramesh, N.S., Gas Loss during Foam Sheet Formation, *Adv. Polym. Tech.*, 15, 297–305, 1996.
115. Weaire, D. and Hutzler, S., *The Physics of Foams*, Oxford University Press, 1999, chap. 13.
116. Ramesh, N.S., Foam Growth in Polymers, in *Foam Extrusion*, Lee, S.T., Ed., CRC Press (ex Technomic), 2000, chap. 5.
117. Benning, C.J., *Plastics Foams Vol. II*, Wiley-Interscience, New York, 1969.
118. Naguib, H.E., Park, C.B., Yoon, E. and Reichelt, N., Fundamental Foaming Mechanisms Governing Volume Expansion of Extruded PP Foams, Foams 2002, Soc. Plas. Eng., 2002.
119. Lee, S.T., Study of Thermoplastic Foam Sheet Formation, *Polym. Eng. Sci.*, 36, 19, 1996.
120. Tobolsky, A.V., *Properties and Structure of Polymers*, John Wiley and Sons, 1960.
121. Yang, C.T., Lee, K. and Lee, S.T., Dimensional Stability of LDPE Foams: Modeling and Experiments, 59th Ann. Tech. Conf., Soc. Plas. Eng., 2001.

122. Hoogendoorn, C.J., Thermal Aging, in *Low Density Cellular Plastics*, Hilyard, N.C. and Cunningham, A., Eds., Chapman & Hall, London, 1994, chap. 6.
123. Burt, J.G., *The Elements of Expansion of Thermoplastics*, 35th Ann. Tech. Conf. (ANTEC), Soc. Plas. Eng. (SPE), 25–28, Apr., 1977.
124. Michaeli, W., Extrusion Dies for the Discharge of a Single Melt, in *Extrusion Dies*, Hanser, Munich, 1984, chap. 5.
125. Thiele, W.C., Foam Extrusion Machinery Features, in *Foam Extrusion*, Lee, S.T., Ed., CRC Press (ex Technomic), 2000, chap. 8.
126. Rauwendaal, C., *Polymer Mixing; A Self-Study Guide*, Hanser, Munich, 1998.
127. Chung, C.I., *Extrusion of Polymers*, Hanser, 2000. chap. 6.
128. Fogarty, J., Fogarty, D., and Rauwendaal, C., Improved Foam Extrusion Output Rates through the Use of Unique Flight Channel Geometry, Foams 2000 Conference, New Jersey, 2000.
129. Chung, C.I. and Barr, R.A., U.S. Pat. 4,405,239, 1983.
130. Yamaguchi, M. and Gogos, C.G., Quantitative Relation Between Shear History and Rheological Properties of LDPE, *Adv. Polym. Tech.*, 20, 4, 261–269, 2001.
131. Lightle, R.D., Sadinski, R.L., and Lincoln, R.M., Vacuum Extrusion System and Method, US Patent 5,753, 161, 1998.
132. Bradley, M.B. and Phillips, E.M., Novel Foamable Polypropylene Polymers, 48th *SPE ANTEC* preprints, 1990.
133. Khemani, K.C., Extruded Polyester Foams, in *Polymeric Foams*, Khemani, K.C., ACS, Ed., Symposium Series 669, 1997, chap. 5.
134. Eckardt, H., Latest Development with Structural Foam, Foam Conference, RAPRA, Heidelberg, Germany, 2002.
135. Pierick, D. and Janisch, R., Molding Technology: Introduction, Applications and Advantages, Foaming Conference, RAPRA, Frankfurt, Germany, 2001.
136. Okamoto, K., *Microcellular Processing*, Hanser, Munich, Germany, 2003.
137. Kiyono, H., Ishimoto, A., Noda, Y., and Yada, K., U.S. Pat. 4,124,344, 1978.
138. Yoshimura, S., Yamaguchi, T., Hashiba, M., and Kanbe, M., Pre-Foamed Particles of Polyethylene Resin, U.S. Patent 4,656,197, 1987.
139. Yamada, H., Hosoda, S. and Ogawa, T., Foamed Particles of Propylene-Type Polymer, U.S. Patent 4,766,157, 1998.
140. Kitamori, Y., PE-PS Polymer Alloy Foam, in *Proc. Thermoplastic Foams Technical Conf.*, Indus. Tech. Reseach Inst., Taipei, Taiwan, ROC, 1995.
141. Reedy, M.E. and Dudek, S., New Polymeric Foam Technologies, Foaming Conference, RAPRA, Frankfurt, Germany, 2001.
142. Sievers, W. and Zimmermann, A., Nanoporous Interpenetrating Organic-Inorganic Networks, U.S. Patent 6,825,260, 2004.
143. Priester, R.D. and Turner, R.B., The Morphology of Flexible Polyurethane Matrix Polymers, in *Low Density Cellular Plastics*, Hilyard, N.C. and Cunningham, A., Eds., Chapman & Hall, London, 1994, chap. 4.
144. Gibson, L.J. and Ashby, M.F., *Cellular Solids: Structure and Properties*, Pergamon Press, Oxford, 1988, chap. 5.
145. Columbo, E.A., *Controlling the Properties of Extruded Polystyrene Foam Sheet*, *Science and Technology of Polymer Processing*, Suh, N.P. and Sung, N., Eds., Massachussetts Institute of Technology Press, Cambridge, MA, 1979.
146. Roberts, A.P. and Garboczi, Elastic Properties of Model Random Three-Dimensional Open-Cell Solids, *J. Mech. Phys. Solids*, 50, 1, 33–55, 2002.
147. Park, C.B., Behravesh, A.H. and Venter, R.D., Low-Density, Microcellular Foam Processing in Extrusion Using CO_2, *Polym. Eng. Sci.*, 38, 11, 1812–1823, 1998.

148. Todd, D. B., Heat Transfer Issues in Foam Extrusion, Foams RETEC.
149. Fogarty, D., Fogarty, J., Rauwendaal, C. and Grald, E., Flow Analysis and Actual Performance Data Turbo-Screw, Foams RETEC, preprints, 55–64, 2002.
150. Ankrah, S., Verdejo, R., and Mills, N.J., Ethylene-Styrene Interpolymer Foam Blends: Mechanical Properties and Sport Applications, *Cell. Polym.*, 21, 4, 2002.
151. Buttler, T., Structure and Properties of Polyolefin Foams, Plastic Foams Conf., Plastics Inst. Amer. at Stevens Institute of Technology, 1973.
152. Glicksman, L.R., Heat Transfer in Foams, in *Low Density Cellular Plastics*, Hilyard, N.C. and Cunningham, A., Eds., Chapman & Hall, London, 1994, chap. 5.
153. Subramonian, S., Burgun, S., Park C.P., Preparation of a Macrocellular Acoustic Foam, U.S. Patent 6,720,363, 2004.
154. Additives, *Modern Plastics Encyclopedia*, 1998.
155. Oshima, M., Polymeric Foaming Simulation: Batch and Continuous, in *Foam Extrusion*, Lee, S.T., Ed., CRC Press (ex Technomic), 2000, chap. 6.
156. Alabak, R.J., Tadmor, Z. and Talmon, Y., Polymer Melt Devolatilization Mechanisms, *AIChE J.*, 36, 9, 1313–1320, 1990.
157. Roberts, A.P. and Garboczi, E.J., *Acta Materiala*, 49, 2, 189–197, 2001.
158. Williams, D.J., *Polymer Science and Engineering*, Prentice-Hall, 1971, chap. 10.
159. Chaudhary, B.I. and Barry, R.P., Extruded Non-Crosslinked Foams from Ethylene-Styrene Interpolymers, Foams RETEC preprints 19–34, New Jersey, 1999.
160. Pontiff, T. and Rapp, J.P., U.S. Patent 5,059,376, 1991.
161. Lee, S.T. and Oiestad, A., Method for Accelerating Removal of Residual Blowing Agent from Extruded Flexible Foams, U.S. Patent 5,411,689, 1995.
162. Kolosowski, P.A., Method for Providing Accelerated Release of a Blowing Agent from a Plastic Foam, U.S. Patent 5,585,058, 1998.
163. Fiddelaers, M. and Swennen, L., Method of Forming Boards of Foam Polyolefin Using Needle Punching to Release Blowing Agent, U.S. Patent 5,776,390, 1998.
164. Lee, S.T., and Lee, K., Solubility of Simple Alkanes in Polyethylene and Its Effects in Low Density Foam Extrusion, Foam Conference, RAPRA, Frankfurt, 2001.
165. Burt, J.G., and Franklin, B.M., Method and Apparatus for Recovering Blowing Agent in Foam Production, U.S. Patent 4,531,951, 1985.
166. Yamazaki, T., U.S. Patent 5,208,266, 1993.
167. Yamaguchi, M. and Suzuki, K., Rheological Properties and Foam Processability for Blends of Linear and Crosslinked Polyethylenes, *J. Polym. Sci. B: Polym. Phys.*, 39, 2159–2167, 2001.
168. Gotoh, S., Fujii, M. and Kitagawn, S., Reactive Polyolefin, *SPE ANTEC*, 54–57, 1986.
169. Throne, J.L., *Understanding Thermoforming*, Hanser, 1999, chap. 2.
170. Suh, K.W., Polystyrene and Structural Foam, in *Polymeric Foams*, Klempner, D. and Frisch, K.C., Eds., Hanser, Munich, 1991, chap. 8.
171. Munstedt, H. and Kurzbeck, S., International Symposium on Elongational Flow of Polymeric Systems, 13–15, Yamagata U., Yonezawa, Japan, 1998.
172. Bradley, M.B. and Phillips, E.M., Novel Foamable PP polymers, *SPE ANTEC Tech. Papers*, 36, 717, 1990.
173. Ratzsch, M., Bucka, H. and Panzer, U., Polypropylene Foams, in *Polypropylene: An A-Z Reference*, Karger-Kocsis J., Ed., Kluwer, Dordrecht, The Netherlands, 1999.

174. Ratzsch, M., Panzer, U., Hesse, A. and Bucka, H., Developemnts in High Melt Strength I-PP: Technology Properties, Applications and Markets, 59th *SPE ANTEC* preprints, 2071–2075, 1999.

175. Naguib, H.E., Park, C.B., Panzer, U. and Reichelt, N., Strategies for Achieving Ultra Low-Density PP Foams, *Polym. Eng. Sci.*, 42, 7, 1481–1492, 2002.

176. Naguib, H.E., Wang, J., Park, C.B., Mukhopadhyay, A.K. and Reichelt, N., Effect of Recycling on the Rheological Properties and Foaming Behaviors of Branched PP, *Cell. Polym.*, 22, 1, 1–22, 2003.

177. Feichtinger, K., Crosslinked Foamable Composition of Silane-Grafted, essentially Linear Polyolefins Blended with PP, U.S. Patent 5,929,129, 1999.

178. Tusim, M.H., Connell, M.C., Suh, K.W., Christenson, C.P. and Park, C.P., Energy Absorbing Articles of Extruded Thermoplastics Foams, U.S. Pat. 6,213,540, 2001.

179. Pfaendner, R., Hoffmann, K., and Herbst, H., U.S. Pat. 5,807,932, 1998.

180. Al Ghatta, H., Giordano, D. and Severini, T., New Developments in PET Foam, Foaming Conference 2001, RAPRA, Frankfurt, 2001.

181. Throne, J., Low-Density PET Foam, Part I, Sherwood Technology Inc., 1997.

182. Brathun, R. and Zingsheim, P., PVC Foams, in *Polymeric Foams*, Klempner, D. and Frisch, K.C., Eds., Hanser, Munich, 1991, chap. 10.

183. Patterson, J., *Vinyl Foam Technology: Trends/New Developments*, 62nd *SPE ANTEC* preprints, 3304–3308, 2002.

184. *Handbook of polymer composites for engineers*, Hollaway, L., Ed., Woodhead Publishing Ltd., Cambridge, U.K., 1994.

185. Storck, W., *Chem. Eng. News*, 65 (44), 15, 2002.

186. Kline & Company, Inc, *6th International Conference on Woodfiber-Plastic Composites*, Madison WI, May 14, 2001.

187. Eller, R., *Plast. World*, 32, 74, 1974.

188. Maldas, D., Kokta, B.V., Raj, R.G., and Sean, S.T., Materials Science and Engineering, 104, 235, 1988.

189. Balatinecz, J.J. and Woodhams, R.T., *J. Forestry*, 91 (11), 22, 1993.

190. Eckert, C.H., *5th International Conference on Woodfiber-Plastic Composite*, pp19, May, 1999.

191. Lu, M., Collier, J.R. and Collier, B.J., *SPE ANTEC Technical Papers*, Vol. I, 1433–1437, Boston, 1995.

192. Richardson, M.O.W., *Polymer Engineering Composites*, Applied Science Publishers, Leicestershire, U.K., 1977, pp. 35–42.

193. Titow, W.V. and Lanham, B.J., Reinforced Thermoplastics, *Applied Science Publishers*, London, 1975.

194. Edenbaum J. and Van Nostrand Reinhold, Eds., *Plastics Additives and Modifiers Handbook*, New York, 1992.

195. Youngquist, J.A. and Rowell, R.M., in *Proc. 23rd International Particleboard/Composite Materials Symposium*, Maloney, T.M., Ed., Washington State University, Pullman, WA, 1990, p. 141.

196. Youngquist, J.A., *Forest Products J.*, 45 (10), 25, 1995.

197. Chtourou, H., Riedl, B. and Ait-Kadi, A., *J. of Reinforced Plastics and Composites*, 11, 372, 1992.

198. Bunsell, A.R., *Fiber Reinforcements for Composite Materials, Vol. 2*, Elsevier, Amsterdam, 1988.

199. Haygreen, J.G. and Bowyer, J.L., *Forest Products and Wood Science: An Introduction*, Iowa State University Press, Ames, IA, 1996.

200. Lewin, M. and Goldstein, I.S., Eds., *Wood Structure and Composition*, Marcel Dekker, Inc., New York, 1991.
201. Patterson, J., *J. Vinyl Additive Tech.*, 7 (3), 138, 2001.
202. Schut, J.H., *Plastics Technology*, July, 2001.
203. Matuana, L.M., Park, C.B. and Balatinecz, J.J., *Polym. Eng. Sci.*, 37 (7), 1137, 1997.
204. Matuana, L.M., Park, C.B. and Balatinecz, J.J., *Cell. Polym.*, 17 (1), 1, 1998.
205. Matuana, L.M., Park, C.B. and Balatinecz, J.J., *Polym. Eng. Sci.*, 38 (11), 1862, 1998.
206. Matuana, L.M., Park, C.B. and Balatinecz, J.J., *Journal of Cellular Plastics*, 32(5), 449, 1996.
207. Park, C.B, Rizvi, G.M., Zhang, H., U.S. Patent No. 6,936,200 B2, 2005.
208. Guo, G., Wang, K.H., Rizvi, G.M., Kim, Y.S. and Park, C.B., PPS-20, Paper #105, Akron, OH, June 20–24, 2004.
209. Park, C.B., Behravesh, A.H. and Venter, R.D., *Polym. Eng. Sci.*, 38 (11), 1812, 1998.
210. Rizvi, G.M., Park, C.B., Lin, W.S., Guo, G. and Pop-Iliev, R., *Polym. Eng. Sci.*, 43 (7), 1347, 2003.
211. Rizvi, G.M., Ph.D. thesis, University of Toronto, 2002.
212. Rizvi, G.M., Pop-Iliev, R. and Park, C.B., *J. Cell. Plast.*, 38 (5), 367, 2002.
213. Guo, G., Rizvi, G.M., Park, C.B. and Lin, W.S., *J. of Appl. Polym. Sci.*, 91, 621, 2004.
214. Reedy, M.F., *Plast. Eng.*, 56, 47, 2000.
215. Mengeloglu, F. and Matuana, L.M., *SPE ANTEC Technical Papers*, 59, 3003, 2001.
216. Matuana, L.M. and Mengeloglu, F., *J. Vinyl. Add. Tech.*, 8 (4), 264, 2002.
217. Li, Q. and Matuana, L.M., *J. Appl. Polym. Sci.*, 88 (14), 3139, 2003.
218. Torres, F.G., Grande, C. and Ochoa, B., *PPS-18*, Guimaraes, Portugal, June 16–20, 2002.
219. Boutillier, P.E., U.S. Patent No. 3,764,642, 1973.
220. Cope, C.W., U.S. Patent No. 5,508,103, 1996.
221. Cope, C.W., U.S. Patent No. 5,847,016, 1998.
222. Rizve, G.M., Matuana, L.M. and Park, C.B., *Polym. Eng. Sci.*, 40 (10), 2124, 2000.
223. Rizvi, G.M., Park, C.B., Guo, G. and Wang, K.H., *SPE ANTEC Technical Papers*, Paper #1041, Nashville, TN, May 4–8, 2003.
224. Naguib, H.E., Park, C.B. and Reichelt, N., *J. Appl. Polym. Sci.*, 91 (4), 2661, 2004.
225. Zhang, H., Rizvi, G.M. and Park, C.B., *Adv. Polym. Tech.*, 23 (4), 263, 2004.
226. Zhang, H., Rizvi, G.M., Lin, W.S., Guo, G. and Park, C.B., *SPE ANTEC Technical Papers*, 47, 1746–1758, Dallas, TX, May 6–10, 2001.
227. Guo, G., Lee, Y.H., Rizvi, G.M. and Park, C.B., *SPE ANTEC Technical Papers*, Paper #102038, Boston, May 1–5, 2005.
228. Bledzki, A.K. and Faruk, O., *SPE ANTEC Technical Papers*, Paper # 154, 2002.
229. Bledzki, A.K. and Faruk, O., *Cellular Polymers*, 21, 417, 2002.
230. Zhang, W. and Bledzki, A.K., *3rd International Wood and Natural Fibre Composites Symposium*, 19-20 Kassel, 33, 1–3, 2000.
231. Bledzki, A.K., Zhang, W. and Faruk, O., *Wood Science and Technology*, 63 (1), 30, 2005.
232. Bledzki, A.K., Gassan, J. and Zhang, W., *J. Cell. Plast.*, 35, 550, 1990.
233. Bledzki, A.K. and Faruk, O., *SPE ANTEC Technical Papers*, 2665–2669, 2004.
234. Winata, H., Turng, L.S., Caulfield, D.F., Kuster, T., Spindler, R. and Jacoson, R., *SPE ANTEC Technical Papers*, 701–705, 2003.
235. Maine, F.W. and Newson, W.R., Patent WO 01/45915 A1.
236. Newson, W.R. and Maine, F.W., *Progress in Woodfibre-Plastic Composites*, May 23-24, Toronto, 2002.

237. Ciferri, A. and Ward, I.M., *Ultra-High Modulus Polymers*, Applied Science Publishers, London, 1979.
238. Zachariades, E. and Porter, R.S., *The Strength and Stiffness of Polymers*, Marcel Dekker, Inc., New York, 1983.
239. Coates, P.D. and Ward, I.M., *Polym. Eng. Sci.*, 21, 612, 1981.
240. Coates, P.D. and Ward, I.M., *J. Mater. Sci.*, 15, 2897, 1980.
241. Gibson, A.G. and Ward, I.M., *J. Mater. Sci.*, 15, 979, 1980.
242. Coates, P.D., Gibson, A.G. and Ward, I.M., *J. Mater. Sci.*, 15, 359, 1980.
243. Kim, Y.S., Guo, G., Wang, K.H., Park, C.B. and Maine, F.W., *SPE ANTEC Technical Papers*, Paper #809, May 16–19, 2004.
244. Kim, Y.S., Guo, G. and Park, C.B., Foams 2004, Wilmington, DE, October 5-6, 2004.
245. Friedrich, K., Application of Fracture Mechanics to Composite Materials, *Composite Materials Series 6*, Elsevier, New York, 1989.
246. Matuana, L.M., Woodhams, R.T., Balatinecz, J.J. and Park, C.B., *Polymer Composites*, 19 (4), 446, 1998.
247. Giannelis, E.P., *Adv. Mater.*, 8, 29, 1996.
248. Wang, K.H., Choi, M.H., Koo, C.M., Choi, Y.S. and Chung, I.J., *Polymer*, 42, 9819, 2001.
249. Okamoto, M., Nam, P.H., Maiti, P., Kotaka, T., Hasegawa, N. and Usuki, A., *Nanoletters*, 1, 295, 2001.
250. Colton, J.S. and Suh, N.P., *Polym. Eng. Sci.*, 27 (7), 493, 1987.
251. Han, X., Zeng, C., Lee, L.J., Kurt, K.W. and Tomasko, D.L., *Polym. Eng. Sci.*, 43, 1261, 2003.
252. Kwak, M., Lee, M. and Lee, B.K., *SPE ANTEC Technical Papers*, 48, Paper #381, 2002.
253. Fujimoto, Y., Ray, S.S., Okamoto, M., Ogami, A., Yamada, K. and Ueda, K., *Macromolecular Rapid Communications*, 24, 457, 2003.
254. Mitsunaga, M., Ito, Y., Ray, S.S., Okamoto, M. and Hironaka, K., *Macromolecular Materials and Engineering*, 288, 543 (2003).
255. Taki, K., Yanagimoto, T., Funami, E., Okamoto, M. and Ohshima, M., *Polym. Eng. Sci.*, 44 (6), 1004, 2004.
256. Di, Y., Iannace, S., Maio, E.D. and Nicolais, L., *J. Polym. Sci.: Part B: Polym.Phys.*, 43, 689, 2005.
257. Shen, J., Zeng, C. and Lee, L.J., Polymer, 46, 5218, 2005.
258. Guo, G., Wang, K.H., Park, C.B., Kim, Y.S. and Li, G., *SPE ANTEC Technical Papers*, Paper #520, May 16–19, 2004.
259. Guo, G., Lee, Y.H., Park, C.B., Kim, Y.S. and Sain, M., Natural Fiber and Wood Composites 2004 Conference, New Orleans, LA, December 8–10, 2004.
260. Gilman, J.W., Jackson, C.L., Morgan, A.B., Harris, R.H., Manias, Jr., E., Giannelis, E.P., Wuthenow, M., Hilton, D. and Philips, S.H., *Chem. Mat.*, 12, 1866, 2000.
261. Finley, M.D., U.S. Patent No. 6054207, 2000.
262. Finley, M.D., U.S. Patent No. 6342172, 2002.
263. Stucky, D.J, Elinski, R., U.S. Patent No. 6,344,268, 2002.
264. Zehner, B.E., U.S. Patent No. 6,590,004, 2003.
265. Strandex Corporation, http://www.strandex.com/licensees.html.
266. Narayan, R., Biobased & Biodegradable Polymer Materials: Rationale, Drivers, and Technology Exemplars, ACS Symposium Ser., in press, 2005.

267. Narayan, R., Drivers for use of biobased and biodegradable polymer materials and emerging technologies, 40th International Symposium on Macromolecules, MACRO 2004, Paris, 2004.

268. Environment Australia, report, Nolan-ITU, Oct., 2002.

269. Ramesh, N.S. and Malwitz, N., SPE ANTEC, 2171–2175, 1995.

270. Lee, S.T. and Ramesh, N.S., *SPE ANTEC*, 3033, 1993.

271. Han, C.D. and Ma, C.Y., *J. Appl. Polym. Sci.*, 28, 831, 1983.

272. Wilke, C.R., *Chem. Eng. Prog.*, 46, 95–104, 1950.

273. Arefmanesh, A. and Advani, S.G., *Rheo, Acta.*, 20, 274, 1991.

274. Ramesh, N.S., Rasmussen, D.H. and Campbell, G.A., *Poly. Eng. Sci.*, 31, 1657, 1991.

275. Bird, R.B. et al., *Dynamics of Polymeric Liquids*, John Wiley & Sons, 1987.

276. Tatarka, P.D., *SPE ANTEC*, 2225, 1995.

277. Nabar, Y., Raquez, J.M., Dubois, P. and Narayan, R., Production of starch foams by twin-screw extrusion: Effect of maleated poly(butylene adipate-co-terephthalate) as a compatibilizer, *Biomacromolecules*, 6 (2), 807–817, 2005.

278. Nabar, Y. and Narayan, R., Twin-Screw Extrusion Production and Characterization of Starch Foam Products for Use in Cushioning and Insulation Applications, *Polym. Eng. Sci.*, in press, 2005.

279. Nabar, Y.U., Draybuck, D. and Narayan, R., Physico-Mechanical and Hydrophobic Properties of Starch Foams Extruded with Different Biodegradable Polymers, *J. App. Polym. Sci.*, in press, 2005.

280. Guan, J. and Hanna, M.A., Post-extrusion steaming of starch acetate foams, *Trans. ASAE*, 46 (6), 1613–1624, Nov.-Dec., 2003.

281. Fang, Q. and Hanna, M.A., Functional properties of polylactic acid starch-based loose-fill packaging foams, *Cereal Chem.*, 77 (6), 779–783, Nov.-Dec., 2000.

282. Fujimoto, Y., Ray, S.S., Okamoto, M., Well-controlled biodegradable nanocomposite foams: From microcellular to nanocellular, *Macrom. Rapid Comm.*, 24 (7), 457–461, May 7, 2003.

283. Hu, X., Nawaby, A.V., Naguib, H.E., Day, M., Ueda, K. and Lia, X., Polylactic Acid (PLA)-CO2 Foams at Sub-critical Conditions, *SPE ANTEC*, 2005.

284. Huneault, M., Bio-Based Polymers: A Review, 19th FoamTech Meeting, IMI NRC-CNRC, 2005.

285. Guan, J. and Hanna, M.A., Selected morphological and functional properties of extruded acetylated starch-Polylactic acid foams, *Ind. Eng. Chem. Res.*, 44 (9), 3106–3115, Apr. 27, 2005.

286. Di, Y.W., Iannace, S. and Di, M.E., Poly(lactic acid)/organoclay nanocomposites: Thermal, rheological properties and foam processing, *J. Polym. Sci., Part B-Polym. Phys.*, 43 (6), 689–698, Mar. 15, 2005.

287. Ray, S.S. and Okamoto, M., Biodegradable polylactide and its nanocomposites: Opening a new dimension for plastics and composites, *Macrom. Rapid Comm.*, 24 (14), 815–840, Sep. 22, 2003.

288. Lu, L., Peter, S.J. and Lyman, M.D., In vitro and in vivo degradation of porous poly(DL-lactic-co-glycolic acid) foams, *Biomaterials*, 21 (18), 1837–1845, Sep., 2000.

289. Nam, Y.S. and Park, T.G., Porous Biodegradable Polymeric Scaffolds prepared by Thermally Induced Phase Separation, *J. Biomed. Mater. Res.*, 47, 8–17, 1999.

290. Hua, F.J., Kim, G.E. and Lee, J.D., Macroporous Poly(L-lactide) Scaffold 1. Preparation of a Macroporous Scaffold by Liquid-Liquid Phase Separation of a PLLA-dioxane-water system, *J. Biomed. Mater. Res.*, 2001.
291. Iannace, S., Di, Y., Maio, E.D. and Marrazzo, C., Biodegradable Foams Based on PCL and PLA: How to Improve the Foaming Process by Means of Reactive Processing and Nanocomposites, Blowing Agents and Foaming Processes, RA-PRA Conference, Germany, 2005.
292. Reedy, M.E., Natural Fiber and Wood Foam Composites—2005, Blowing Agents and Foaming Processes, RAPRA Conference, Germany, 2005.

Author Index

Subject Index

A

Acrylonitrile-butadiene-styrene (ABS), 134
Activation energy, 33, 36, 160, 176, 192
Additives, 40, 44, 73, 78, 104, 109, 111
    high amylose and hydroxypropylated
        starch blend foam, 185, 186
    PLA, 189
    starch foam, 182
    wood fiber composites, 135
Adipic acid, 167
Aerogel, 91–92
Aging, 43, 74, 90
    diffusion coefficient, 39
    PE foam dimensional variation, 108–111
    P-V-T variation, 70, 71
    residual blowing agent, 112–113
    thermoplastic foam, 14
Air flow, 109
Aliphatic-aromatic copolyesters (AAC), 188
Aliphatic polyesters (ALP), 188
Alumina trihydrate, 132
Andersen Corporation, 163
Anhydro-D-glucopyranose, 133
Argon, 36, 42
Aromatic polyesters (ARP), 188
ASTM D-2586-A method, 97
ASTM D-635, 161
ASTM D-6400, 170
Automotive, 1, 21, 77, 87, 99, 117–119,
        126–129, 133, 135, 182–183
Avrami equation, 36
Azodicarbonamide, 30

B

Bacterial, 122, 168, 170
Bamboo, 133
Barrier properties, 157
Batch foaming, 54, 143
Biodegradable, 20, 49, 92, 165–192
Biomass, 166, 167, 170

Bio-organics, 166
Biopolymers, 167
Block copolymers, PP with ethylene content,
    127
Blowing agents, 10–13
    chemical, 43–45 (*See also* Chemical
        blowing agents (CBAs))
    explosive limits, 114
    fundamental features, 12
    physical, 41–43 (*See also* Physical blowing
        agents (PBAs))
    property considerations, 13
    residual, 113
    volatility, 42
Brabender, 149, 151
Branching, 129
    effect on melt strength, 66
Bubble growth, 60

C

Calcium carbonate, 132
Capillary effects, 122
Capillary rheometers, 176
Carbon dioxide, 19, 29, 41, 42, 90, 111, 113,
        121, 143, 168, 170, 190
    amount dissolved in polyethylene, 59
    increase with increased amount of water,
        19
    limited solubility and dissolution, 48
Carpet backing, 117
Cavitation, 54
Cavity nucleation, 58
Cell(s)
    breakage, 65
    closed, 48, 49, 66, 95
        under compression, 101
        moisture absorption, 129
        open cell vs., 9, 94
    coalescence, 18, 52, 74, 105, 106, 137
        rupture and, 64–66
    cubic model, 94, 95, 109